The Green Graduate

The Green Graduate
Educating Every Student as
a Sustainable Practitioner

SAMUEL MANN

NZCER PRESS
Wellington 2011

NZCER PRESS
New Zealand Council for Educational Research
PO Box 3237
Wellington
New Zealand

© Samuel Mann, 2011

ISBN 978-1-877398-99-5

All rights reserved

Designed by Cluster Creative

Printed by Printlink

This title is also available as an e-book
from www.nzcer.org.nz/nzcerpress

Printed on paper manufactured under the
ISO14001 Environmental Management System
from elemental-chlorine-free, Forest Stewardship
Council-certified Mixed Source pulp.

Distributed by NZCER Distribution Services
PO Box 3237
Wellington
New Zealand
www.nzcer.org.nz

Contents

Acknowledgements		9
PART 1 INTEGRATING SUSTAINABILITY INTO ACADEMIC PROGRAMMES		11
1	**Introduction**	13
1.1	How to use this book	14
1.2	A framework for sustainable practice	16
1.3	The structure of this book	18
2	**The Otago Polytechnic journey**	19
2.1	The objectives	21
2.2	The framework in practice	24
3	**Education for sustainability**	29
3.1	The biggest levers	33
3.2	The role for higher education	34
3.3	Every graduate	35
3.4	The sustainable practitioner	36
4	**Transformation: Process and curriculum together**	45
4.1	Separate course or integration?	46
4.2	Sustainability curricula	51
4.3	Systems and systemic thinking	53
4.4	Critical and creative thinking	59
4.5	Ethics	60
4.6	Willingness to act	61
5	**Transformation**	65
5.1	Journey, not destination	66
5.2	Top down or bottom up?	68
5.3	Telling the story	69
5.4	A new role for educators	72
5.5	Footprint and handprint	72

6	**Sustainability in IT and computing**	**77**
6.1	Incoming student survey	77
6.2	Curriculum initiatives	80
6.3	SimPā	87

7	**Sustainability in other disciplines**	**91**
7.1	Art	91
7.2	Design	94
7.3	Midwifery	95
7.4	Nursing	96
7.5	Occupational therapy	97
7.6	Sport	99
7.7	Veterinary nursing	100

PART 2 BARRIERS TO INTEGRATING SUSTAINABILITY INTO ACADEMIC PROGRAMMES — **103**

8	**An overview of barriers**	**105**
8.1	Identifying and overcoming barriers: The theory	106
8.2	What we know about barriers to EfS	108
8.3	Structure of barriers and responses	111

9	**Is there really a problem, and is sustainability the answer?**	**113**
9.1	Ecosystems are not really under stress and declining, affecting our future	113
9.2	If we do need to act, we've got a while yet	115
9.3	Is sustainability the answer?	117
9.4	Do we even understand what sustainability means?	119
9.5	It's their problem, not mine	121
9.6	But what about the bottom line?	122
9.7	This stuff is just leftist nonsense	124

10	**Is it the responsibility of higher education?**	**127**
10.1	Is it really up to us to do this?	127
10.2	Shouldn't higher education be values free?	131
10.3	What does it mean for links with community and business?	132
10.4	Is this what students sign up for?	136
10.5	Surely it's the role of the media rather than of higher education	136
10.6	They don't need this, it's all over the Web	137
11	**I can't see this working in our institution**	**139**
11.1	What happened to academic freedom?	139
11.2	What about the sustainability of our student intake?	140
11.3	Sustainability is not recognised by our funding model	143
11.4	Sustainability is not valued by my institutional hierarchy	143
11.5	How does this fit with all we're required to do?	144
11.6	We'd rather the students were critical thinkers	146
11.7	We're doing it already in our own way	146
12	**I want to integrate EfS, but how do I do it?**	**149**
12.1	I can't really see how it fits my discipline	149
12.2	It seems such a big job to do it properly	151
12.3	The professional development didn't help much	152
12.4	Isn't it a bit hypocritical not to practise what we preach?	153
12.5	Where do I start? It isn't in the texts I use	154
12.6	I'm not confident that I have the expertise	155
12.7	I don't want to preach to the students	157
12.8	How do I accommodate all the different views and perspectives?	158

Conclusion	**161**
References	**163**
The Beeby Fellowship	**175**

Acknowledgements

This book has benefited greatly from many people's ideas, passion and toil.

Special credit goes to the staff and students of Otago Polytechnic. In particular I would like to acknowledge the outstanding leadership of Phil Ker—thank you for recognising the potential for sustainability to transform the institution. My debts to colleagues across the institution are numerous: Dave Bremer, Nicola Bould, Robin Day, Katie Ellwood, Steve Henry, Anna Hughes, Mark Jackson, Barry Law, Ella Lawson, Glenice Mayo, Alistair Regan, Michelle Ritchie, Khyla Russell (and SimPā team), Lesley Smith, Anni Watkins and Chris Williamson. Of particular note are the Sustainability Champions across the institution, and the staff and students of the School of Information Technology. I should also like to acknowledge the Otago Polytechnic Council, past and present, for its courage in positioning sustainability as a cornerstone of Inspiring Capability. The members of Industry Liaison Committees and other local and national professional bodies deserve similar mention for their patience and vision in embracing a sustainable future for their trade or profession.

I should also like to acknowledge the professional companionship on this journey of Nell Buissink-Smith, Peter Holland, Kate Kearins, Richard Morgan, Hayden Montgomerie, Logan Muller, Kerry Shephard, Pam Williams, and colleagues from NACCQ/CITRENZ, especially Alison Young, Tony Clear, Garry Roberton and Steve Corich. The assistance of Richard West is hugely appreciated, as is the work of those at NZCER Press: Bev Webber, Helena Barwick and Joanna Morton.

In the spirit of the link across generations, this book is for my parents, and for my children: Phoebe, Oliver and Henry.

PART ONE
INTEGRATING SUSTAINABILITY INTO ACADEMIC PROGRAMMES

CHAPTER ONE

Introduction

This book explores Otago Polytechnic's commitment to the view that every graduate should be able to think and act as a *sustainable practitioner*.

My business card says that I'm an Associate Professor in Information Technology at the polytechnic, and I hold a portfolio in Education for Sustainability. Here's the first question that people always ask me: "You're in computing—that's not sustainability, how come you're the sustainability guy?" I used to answer, "Oh, I used to be a botanist/geographer," which kept everyone happy but had me increasingly thinking. If I was a botanist, no one would have asked that question (and I wouldn't have had such a ready answer!). Those thoughts coalesced into the key principle behind the approach adopted at Otago Polytechnic. I should have said something like:

> That's a good question, but my hope is that in a few years you wouldn't ask it. Sustainability is something that affects all of us, in every discipline, no less a computer scientist than a botanist, no less a nurse than a geographer.

Whatever discipline we might be from, we can ask ourselves what our trade or profession can contribute to a sustainable future—how people in our industry can act as sustainable practitioners.

From this viewpoint, every graduate needs to develop sustainability competencies. Using the Otago Polytechnic story as a case study, this book focuses on how sustainability can be woven through teaching and learning across all discipline areas. It explores what it means to prepare computer scientists, botanists, nurses and geographers to contribute to a sustainable future. I argue that rather than having a separate course—an "Education for Sustainability 101"—the aim should be for elements of sustainability to be incorporated into all disciplines and learning experiences and for sustainability to be an integrated core competency for all training outcomes.

The goal is to ensure that sustainability becomes part of business as usual, even if business as usual is different for each graduate. The approach, commonly called "every graduate", means just that: sustainability is not an optional extra or something for a few experts or heroes, but something that is integrated into every programme.

Sustainability is a journey rather than a destination.[1] We will probably never get there—there will always be something we can do better. The real measure of our success will be when graduates are empowered to do the right thing in the workplace.

1.1 How to use this book

This book is designed to be used in a number of ways. How you use it, and what you get from it, might depend on where you stand on the teaching of sustainability. In Box 1 there are a number of statements. See how many of them you agree with. Most people agree with everything up to the last statement: "I am currently integrating sustainable practice into my work (i.e., acting as a sustainable practitioner)."

---

1 Compare Roy Goodman's "Remember that happiness is a way of travel, not a destination", after Ralph Waldo Emerson's "Life is a journey, not a destination".

INTRODUCTION

> **BOX 1**
>
> Note when you stop agreeing with these statements:
>
> 1. Ecosystems are under stress and declining, and this is affecting human conditions and futures.
> 2. Sustainability—defined broadly as meeting the needs of all current and future generations—is a reasonable approach to addressing this decline.
> 3. Sustainability is the responsibility of everyone, during their whole lives, including at work.
> 4. This work component—here called "sustainable practice"—applies to every career, every discipline.
> 5. Acting as a sustainable practitioner means both reducing my footprint (reducing harm) and increasing my handprint (actions towards sustainability).
> 6. I am currently integrating sustainable practice into my work (i.e., acting as a sustainable practitioner).

For educators, the last statement leads to four more statements that describe the educator's role in preparing graduates for their careers. These statements are in Box 2. Do you agree with them? Again, most people agree with the statements until the very last one: "I am currently integrating education for sustainability into my teaching."

What this tells us is that people recognise that sustainability is important, they know they should be doing something about it and they understand this means integrating Education for Sustainability (EfS) into their teaching. That they are not doing so suggests there are barriers to be overcome.

There are a lot of books about personal sustainability, and some about sustainable practice. However, there are few about how to prepare people to be sustainable practitioners. This book aims to fill that gap.

The subtitle of the book is *Educating Every Student as a Sustainable Practitioner*, and that sums up its approach. It is about **education**—the

> **BOX 2**
>
> Note when you stop agreeing with these statements:
>
> 1. It is the responsibility of higher education to, among other things, prepare every graduate for their careers as sustainable practitioners.
> 2. Preparing people for their careers as sustainable practitioners is part of the responsibility of every academic.
> 3. I would like to be able to integrate education for sustainability into my teaching.
> 4. I am currently integrating education for sustainability into my teaching.

focus is on teaching and learning, not recycling or the other things that we commonly associate with operational sustainability. It is about **every student**—that is, every student in every discipline, not just those who are interested in sustainability or are studying in areas where it might seem more relevant. And it is about preparing these students as **sustainable practitioners**—providing them with the skills, values and behaviours they will need in order to contribute to a sustainable future.

1.2 A framework for sustainable practice

The framework for sustainable practice discussed here is both top down and bottom up. It is based on the concept of the sustainable practitioner and requires each discipline to work with its community, along with professional and industry bodies, to identify what it means to be a sustainable practitioner in that area. In this way, teaching practices are directly relevant to community and industry needs. The components of the framework are summarised below, and discussed in more detail later.

INTRODUCTION

Identify what it means to be a **sustainable practitioner (SP) for each discipline**, which means:

- working with the wider community to envisage and articulate a role for each discipline's practitioners for a sustainable future
- working with the wider community to articulate a discipline's response to sustainability—this may take the form of mission statements (etc.) from professional societies
- working with the wider community to identify expected behaviours for practitioners when confronting sustainability challenges
- developing an understanding of the current status of sustainability (values, awareness, knowledge, skills and behaviours) among all our stakeholders (students, stakeholders, staff, graduates, professional/trade connections and our respective iwi partners).

Make sustainability part of the **graduate profile (GP)**, which means:

- identifying sustainability statements for the graduate profile and core competencies for graduate practitioners—in some cases this is an incremental adjustment and in others a transformative rewrite
- integrating learning outcomes into courses, simultaneously covering both course-specific issues and holistic approaches.

Incorporate sustainability into **student learning (SL)**, which means:

- focusing on relevant and engaging courses
- identifying and promoting exemplar resources and teaching strategies—this includes the pre-existing knowledge of sustainable practices and aspirations for iwi Māori locally
- identifying and addressing sustainability-related areas missing from current curricula
- assessing lecturer expertise and skill requirements in EfS, and establishing a development plan for the department.

 Take a **multidisciplinary (MD)** approach, which means:

- undertaking activities to raise the awareness of sustainability in your discipline, and in particular, making use of multidisciplinary and experiential approaches
- integrating sustainability into quality assurance processes such as curriculum documents, moderation and monitoring checklists
- framing "for sustainability" as a core driver for research—this means research aimed at increasing the sustainability of the discipline and promoting the discipline for sustainability, both of which will require a wider interdisciplinary approach to research
- establishing a network of sustainability champions to promote EfS as a legitimate and mainstream part of the discipline.

1.3 The structure of this book

The book is divided into two parts. Following this introduction, Chapter 2 tells the story of Otago Polytechnic's sustainability journey. Chapter 3 then outlines the history and some of the challenges of EfS. Chapters 4 and 5 discuss what was required to launch Otago Polytechnic on its journey towards making "Every graduate a sustainable practitioner". Chapter 6 describes how the discipline of IT and computing has approached sustainability, and Chapter 7 gives some examples from the other disciplines.

In Part 2 I examine a number of barriers to teaching EfS, categorised under the following headings:
- Is there really a problem, and is sustainability the answer?
- Is it the responsibility of higher education?
- I can't see this working in our institution.
- I want to integrate EfS, but how do I do it?

CHAPTER TWO

The Otago Polytechnic journey

Wals and Jickling (2002) argue that EfS requires creativity: there are no recipes. The recipe metaphor is inappropriate because it treats EfS like baking: get the mix right and out pops a perfectly formed cake. In education we know that reliance on a recipe is doomed to failure, and in sustainability a recipe model is even less helpful. Instead, travel is a more useful model, in particular the idea that sustainability is a journey, not a destination. From this perspective, the framework presented in Chapter 1 can be seen as a map and a set of guiding principles. Happily, there are fellow travellers on the journey.

There are many drivers for a focus on EfS, both local and international. For a start, this is the United Nations' Decade of Education for Sustainable Development, and it is clearly signalled in New Zealand's Tertiary Education Strategy. Even more importantly, though, it is the right thing to do.

In 2007 Otago Polytechnic set itself the goal that "every graduate may think and act as a sustainable practitioner". This goal was aimed at contributing to a better community, at producing graduates across the institution with relevant skills and values, and at working closely with industry to both identify and achieve sustainable practice in each discipline. I look at this goal in more detail in Section 2.1.

Otago Polytechnic's initiative to make every graduate a sustainable practitioner is based on the goal of incorporating EfS across all disciplines and programmes as an integrated core competency for all graduates. This educational integration is based on the idea that because sustainability should be integrated into all activities of daily life, it should also be part of tertiary and vocational training. This is the basis of the "every graduate" approach, with the aim of preparing every graduate of the institution with EfS relevant to their field.

Otago Polytechnic's commitment to EfS has several strands. Figure 1 represents this as a network. The nodes in the network are students, staff, operations etc. The strands are the lines joining those nodes. All the strands that connect to "Students" are part of the learning experience. So while we might focus on transforming

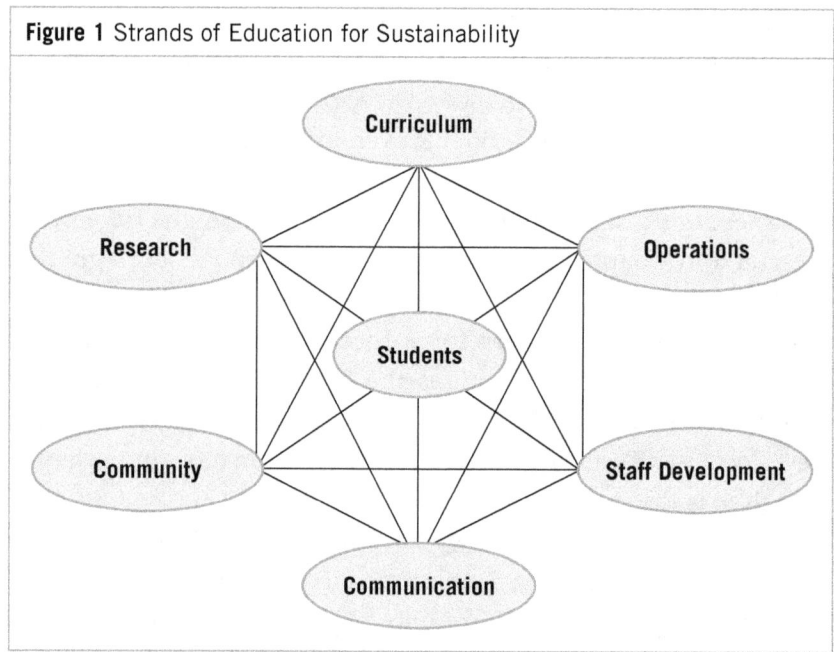

Figure 1 Strands of Education for Sustainability

the predominant part of the learning experience—the link between staff and students—the other lines are vitally important. These other strands provide the "hidden curriculum", demonstrating that we endeavour to "walk the talk". For example, initiatives in Operations include waste and carbon audits, recycling, polybikes, fleet management and supplier agreements. Crucially, all of these things are done with a double intention—both the inherent benefit and the opportunity to reinforce the sustainability message.

Like a commitment to developing Total Quality Management, EfS is an ongoing project. It is therefore important not to forget that Otago Polytechnic is on a journey, and is not making any claim that all is perfect, or that it is a sustainable organisation.

2.1 The objectives

The objectives for Otago Polytechnic's journey are best encapsulated in the Academic Board's statement on sustainability, passed in August 2007. This was not the start of the journey, but it was an important milestone, and one that took considerable consultation with the community and across the campus:

> The skills and values of Otago Polytechnic graduates contribute to every sector of society. Our curriculum, teaching and learning therefore is pervasive and influential with global impact. The Otago Polytechnic sustainability vision is that our graduates, our practitioners and our academics understand the concepts of social, environmental and economic sustainability in order for them to evaluate, question and discuss their role in the world and to enable them to make changes where and when appropriate. Our goal is that every graduate may think and act as a 'sustainable practitioner'.
>
> Moreover, educators must take a lead in sustainability so that our graduates can be encouraged and supported to promote sustainable practices in their chosen career. This can primarily be achieved by fostering education for sustainability in all our qualifications and by

re-visioning and changing our approach to teaching and learning to model a transformative context for all learners.

As a consequence sustainable practice becomes a context and a process for learning and recognised as a core capability within each discipline.

Creating a philosophy of Education for Sustainability will be enhanced if undertaken within a context of institutional operational practice. We will then be seen to be modelling good practice. (Otago Polytechnic, 2007, p. 1)

This statement benefits from unpacking:

The Otago Polytechnic sustainability vision is that our graduates, our practitioners and our academics understand the concepts of social, environmental and economic sustainability in order for them to evaluate, question and discuss their role in the world and to enable them to make changes where and when appropriate.

Otago Polytechnic's approach to sustainability is inclusive in terms of people and concept. We loosely define sustainability as having three intertwined components—social, environmental and economic sustainability. What we would like our people to be able to do is to think creatively and critically and be able to make transformative changes.

Our goal is that every graduate may think and act as a 'sustainable practitioner'.

This is the critical sentence. It sets the goal and states our "every graduate" approach. This means that students in every discipline and at every level of education at Otago Polytechnic should have these characteristics. Note that "may think and act" was very carefully worded: as much as we might have liked to use the word *will*, an educational institution is not in a position to prescribe behaviours following graduation. "Think and act" also highlights the balance between cognitive and action capability (Jensen & Schnack, 1997).

Moreover, educators must take a lead in sustainability so that our graduates can be encouraged and supported to promote sustainable practices in their chosen career.

This important statement sets out our position that sustainability will be a part of careers, and that the institution has a role in promoting such values.

> This can primarily be achieved by fostering education for sustainability in all our qualifications and by re-visioning and changing our approach to teaching and learning to model a transformative context for all learners.

In other words, it is not possible to provide opportunities for transformation without a fundamental examination of both what we teach and how we teach.

> As a consequence sustainable practice becomes a context and a process for learning and recognised as a core capability within each discipline.

Recognising sustainability as a core capability places it at the same level as other generic competencies, such as literacy and numeracy (or, as we are fond of expressing it: "reading, writing and sustainability").

> Creating a philosophy of Education for Sustainability will be enhanced if undertaken within a context of institutional operational practice. We will then be seen to be modelling good practice.

If we are to achieve our goal for the students we teach, the institution needs to be an exemplar of sustainable practice. The decision was therefore made to integrate sustainability education into every programme rather than develop a stand-alone course. This demonstrates Otago Polytechnic's belief that the goal of sustainability in the world will only be achieved through everyone learning to live and work sustainably. In the next section I look at how this has worked in practice.

2.2 The framework in practice

Sustainable practitioner

The concept of the sustainable practitioner is fundamental to our approach. Each discipline is coming to terms with what it means to be a sustainable practitioner. This is expressed as a statement beginning, "A sustainable practitioner in [discipline] is someone who ...", accompanied by a short narrative that describes the desired behaviours.

Each department has worked with its stakeholders to identify the behaviours expected of their own graduates. A comprehensive survey of all incoming students has set a baseline for this initiative. In the long term this will provide evidence of the effectiveness of our approach; in the short term it is highlighting the needs of different student groups—ensuring the teaching of EfS is relevant to each group.

Graduate profile

Sustainability statements for graduate outcomes and core competencies for graduate practitioners have been completed. Generic graduate profiles and learning outcome statements have been developed.

Key sustainability curriculum areas (e.g., systems thinking, ethics, behavioural strategies) are linked with New Zealand Qualifications Authority level descriptors to produce exemplar learning outcomes for education for sustainability.

Student learning

Many transformations have occurred throughout the institution. The initiatives of individual staff and students are being supported through the activities of a Sustainable Learning Co-ordinator. A Living Campus initiative aims to provide integrated learning opportunities through the development of a community garden (an interactive, open-air experience), along with enhancing the sustainability of the campus.

The Living Campus project melds all the components of the sustainability initiative through the development of an interactive sustainability museum and education programmes within a vibrant, productive, community garden. It is envisaged that the Living Campus will become the hub for sustainability-oriented community education services. Staff and students from across the institution are collaborating with the wider community to convert the entire campus into a productive community garden that is also a focus of community engagement.

EfS is also being integrated into curriculum development and moderation processes.

Figure 2 Staff, students and families lay out a no-dig garden during the first of many working bees on the Living Campus

Multidisciplinary Approach

In association with the student union, all student programme handbooks have several pages dedicated to sustainability.

The Sustainable Habitat Challenge is a national collaborative project for teams around New Zealand to design, develop and build sustainable housing in their local community. Otago Polytechnic hosted and organised the 2009 challenge and had two teams involved.

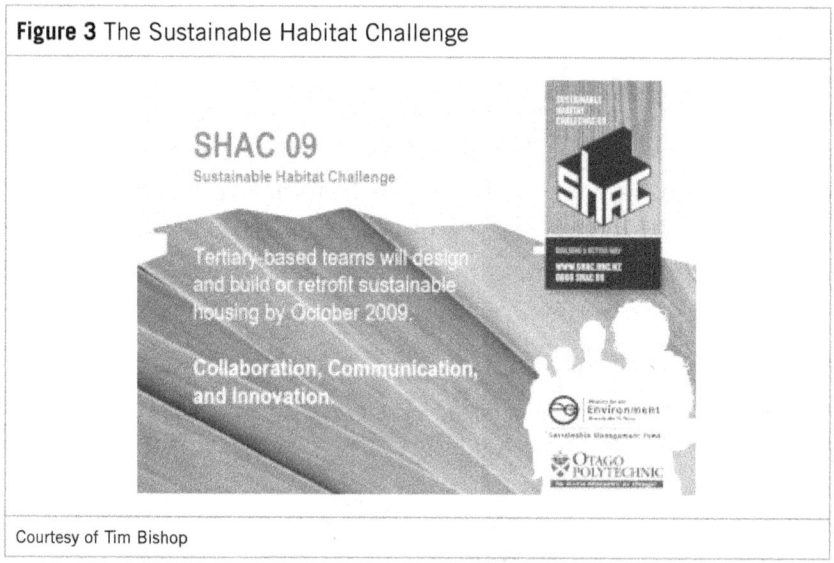

Figure 3 The Sustainable Habitat Challenge

Courtesy of Tim Bishop

A Centre for Sustainable Practice has been established to facilitate community and business engagement in EfS. Activities have included workshops on carbon trading, sustainable tourism business courses, sustainable house tours, sustainable tourism advisers, strategic planning and futures forums.

"What's Best Expos" have been organised by Otago Polytechnic's Centre for Sustainable Practice. These expos provide a vehicle for progressive companies to show their wares, and also to learn from each other.

Several programmes are being developed to meet the needs of people who want to specialise in sustainability. These professional development courses, delivered in flexible formats, include courses to incorporate sustainable management in different industries.

The introduction of these sustainability initiatives has seen a continuum of approaches to teaching sustainability across the campus, from EfS101 to full integration of sustainability concepts in existing courses. Both ends of the spectrum are acceptable, as teaching staff interpret the concepts of sustainability in ways that complement their curriculum area.

For many lecturers and schools this has been a validation of what they are already doing, while for others it has been a challenge to reframe existing courses to take sustainability concepts into account. Most progress has been made in areas where sustainability is seen as a core principle, not an add-on, and where sustainability is viewed as a context for learning.

The approach has been one of empowerment—supporting people who are transforming their teaching practice. To this end, institutional support in areas such as adding footnotes to moderation checklists has been vital; for example, "All Otago Polytechnic assessments may include elements reflecting our commitment to the generic capabilities of literacy, numeracy, sustainability …"

Focusing on the role of the sustainable practitioner within each discipline is fundamental to the Otago Polytechnic approach. Starting by recognising the contribution of the discipline to an improved future has meant we have had almost no resistance from any academic area. The anticipated "This is nothing to do with us" response has not eventuated. Instead, the responses have been more along the lines of, "How can we contribute?", "This is so important to our area" and "At last, someone is listening."

Having briefly looked at how sustainability has been integrated into the teaching at Otago Polytechnic, I want to step back and look at the origins and thinking behind sustainability. I will continue to use Otago Polytechnic to illustrate the various concepts and show how they can be applied in order to provide a fuller picture.

CHAPTER THREE

Education for sustainability

If you are reading this book in the hope of becoming a convert to sustainability then you are reading the wrong book. In writing this book I have made the conscious decision not to try to convince people of sustainability's worth. This is on the basis that the choir doesn't need preaching to and disbelievers aren't listening anyway. I would rather present the action-based sustainable practitioner as a sensible approach for the future of any discipline. But (and to continue the religious analogy), I am increasingly realising that there is a whole bunch of people who don't fit into either the choir or the disbeliever category. I have had several experiences of someone coming up at the end of a development session and saying, "Yes, I get that, but what has it got to do with my teaching?" This chapter is for those people.

Sustainability has a long history. So does unsustainability. Diamond (2005) analyses the factors behind the collapse of societies, both ancient and modern, and finds the fundamental relationship is the one between a society and its climate and geography, resources and neighbours. Societies fail, he concludes when they mismanage the ecosystem—soil, trees and water. This observation of the close relationship between people and the environment is not new. In 1864 George Marsh wrote *Man and Nature; or Physical Geography as Modified by Human Action*. In it he wrote, "Man has too long forgotten that the earth was given to him for usufruct alone, not for consumption, still less for profligate waste" (p. 43).

The legal term *usufruct* is appropriate; it refers to the right to use and benefit from a resource, but not to damage or alienate it. Before Marsh, people knew that humans modify the natural world by cutting forests, draining swamps and depleting wildlife, but these actions were seen as progress. Marsh set out to show the negative side of such human impacts:

> The earth is fast becoming an unfit home for its noblest inhabitant, and another era of equal human crime and human improvidence, and of like duration with that through which traces of that crime and that improvidence extend, would reduce it to such a condition of impoverished productiveness, of shattered surface, of climatic excess, as to threaten the depravation, barbarism, and perhaps even extinction of the species. (p. 43)

At about the same time, John Ruskin observed that the same economic system that creates glittering wealth also spawns what he called "illth"—poverty, pollution, despair, illness. It makes life comfortable for some, but it does so at considerable discomfort to others. This era of intense, irreversible human influence on the Earth's systems has been named the "Anthropocene", and has been defined as that time which began in the early 1800s with the onset of industrialisation (Crutzen, 2002, 2005; Steffen, Crutzen, & McNeill, 2007).

The concept of sustainability has deep roots. The Age of Discovery, the Wars of Religion and the mercantile expansions of the 15th and 16th centuries saw a massive growth in ship building across Europe. This resulted in rapid exploitation of domestic natural resources, especially forest resources, which led to widespread shortages of wood and an acute political problem. Hanns Carl von Carlowitz's 1713 *Silvicultura Oeconomica* (as cited in McDonald & Lassoie, 1996) is considered to be the first text to take a sustainability approach. In it, von Carlowitz expresses the view that logging and reforestation have to be in balance. Heinrich von Cotta's *Anweisung zum Waldbau* (1817/2000) is another cornerstone of sustainability. In it he exposes the essential fallacy of forestry—that human intervention improves forests:

> Here and there we admire still the giant oaks and firs, which grew up without any care, while we are perfectly persuaded that we shall never in the same places be able, with any art or care, to reproduce similar trees. (von Cotta, p. 27)

Von Cotta also introduced some of the core planks of sustainability: answers are complicated by time and space, and by human perception. Variation in systems means that "what many declare good or bad, proves, good or bad only in certain places". He paraphrases Wieland's earlier (1768) statement that "they cannot see the forest for the trees" (as cited in Bartlett, 1980, p. 380) to introduce issues of perceptions of scale. He also suggested humility, both as a society and on a professional level, and that foresters should not assume that they have perfect understanding.

In addition to highlighting mankind's role as the cause of environmental degradation, Marsh (1864, p. 35) also offered advice on the "restoration of disturbed harmonies". He promoted the need for a different sort of relationship between man and nature, whereby man is to become a co-worker with nature in the reconstruction of the damaged environment. From these early conceptions, and with boosts from Rachel Carson's *Silent Spring* (1962) and underlined by the Club of Rome's *The Limits to Growth* (Meadows, Meadows, Randers, & Behrens, 1972), there has been an increasing awareness that human use of the Earth is approaching a range of environmental and resource limits, and that this, rather than diminishing, is escalating at an alarming rate.

Definitions of sustainability vary widely, from a strong environmental focus:

> Sustainable development—improving the quality of human life while living within the carrying capacity of supporting ecosystems (International Union for Conservation of Nature, United Nations Environment Programme & World Wildlife Fund, 1991, p. 10)

through economic approaches:

> Sustainable development: The amount of consumption that can be sustained indefinitely without degrading capital stocks, including natural capital stocks (Costanza, 1991, p. 8)

to a broad global view:

> Sustainable development is development that meets the needs of the present without compromising the ability of future generations to meet their own needs. (World Commission on Environment and Development, 1987, p. 43)

A key feature of the World Commission on Environment and Development definition is its potential to integrate environmental and economic concerns, along with a concern for the wellbeing of all. In this conception, sustainable development implies greater equity and continued growth, but growth of a more environmentally, socially and economically sustainable kind.

We live in a world in which our economic and social systems are interdependent, and both of these systems are completely dependent on the ecological systems upon which all life and all social and economic activity depend. This dependence is reflected in the Strong Sustainability model (Daly, 1996). As a result, current definitions of sustainability generally include three components: environment, society and economy, along with the recognition that these three areas are intertwined (McKeown, 2002).

"Sustainability" and "sustainable development" are often used interchangeably, but they do mean different things. Sustainability usually means the capacity for continuance into the long-term future. Sustainable development is the journey or means of achieving the goal of sustainability. It is easy to become negative about sustainability. To do so, however, is to miss the point. The focus of sustainability is on the solutions, not the problems.

3.1 The biggest levers

Schendler (2009, p. 19) describes the scale of the climate problem as being "so great, that to many people it is incomprehensible". He describes how individual actions such as using canvas bags at the supermarket are necessary but inadequate, saying "we can't afford the delusion that such individual action is enough" (p. 19). Instead, he argues that we should indeed change our light bulbs, but more important is the task of figuring out how we can help ensure that everyone on the planet changes their light bulbs too. Schendler argues:

> To lessen the charges of hypocrisy that could be brought against any of us, it seems obvious that the best thing to do would be to implement even more sustainable practices—the real ones, things that really matter and drive real change. To do that you need to be clear-eyed about how you can make a real difference: *you need to find your biggest lever and use it* [italics added]. (2009, p. 85)

For educators, our biggest lever is our ability to positively influence the skills, values and behaviours of our graduates. This leverage extends beyond sustainability. It is, of course, why we have an education system at all. AtKisson (2008) argues that "change agents—people dedicated to promoting sustainability ideas and innovations—are needed in every field, in ever increasing numbers" (p. 307). He says:

> If the job of sustainable development is to save the world, and if the engagement of increasing numbers of people is required to accomplish

this enormous but possible task, then we have a duty to help as many people as possible to get involved, to help them get better at it—and to survive and thrive economically at the same time. (p. 107)

To achieve this, AtKisson argues that the special role for educators is to "prepare current and future generations for a great responsibility, directing human development towards sustainability, and beyond" (p. 307).

3.2 The role for higher education

Corcoran and Wals (2004) put the imperative for education like this:

> The higher education community is called to respond to times of disastrous anthropogenic environmental crises, failing political systems, religious intolerance, and unsustainable and inequitable economic development. (p. 3)

Similarly, Cortese (1999) explains that society has granted higher education a unique role, in that it receives public and private funds in exchange for educating students and producing the knowledge that will lead to a thriving civil society. Cortese goes on to suggest that current higher education systems are largely failing to help students develop the skills and explore the values, perspectives and knowledge for the deeper learning necessary for sustainable development.

More than just missing the opportunity to prepare sustainable practitioners, several authors argue that education is complicit in *un*-sustainability. Corcoran and Wals (2004) have the view that "the scope and range of the negative impacts of university-educated people on the natural systems that sustain Earth are unprecedented" (p. 3). Lautensach (2004, p. 2) makes a similar point:

> The fact that a large number of well-educated people continue to make decisions that are blatantly counterproductive in the present situation indicates that merely being informed about the crisis does not by itself dispose a person toward responsible behaviour.

Stephens, Hernandez, Roman, Graham and Scholz (2008) examine the role of tertiary education as a change agent. They argue that higher education has a "unique potential to catalyze and/or accelerate a societal transition toward sustainability" (p. 320). In addition to the operational level of change though curriculum, campus operations and research, they recognise higher education can have impact at the strategic and tactical levels. At the strategic level, higher education can be involved in defining and developing strategic societal visioning and setting long-term goals. Fostering and facilitating coalitions and co-operation among stakeholders are considered tactical.

All three of these levels are important in articulating the critical role of higher education in sustainability. Cortese (1999; Cortese & McDonough, 2001) has stressed the importance of articulating the future (he uses the year 2050): as professionals in our disciplines we need to examine what role we see our graduates playing in that future. As educators charged with creating those graduates, we are *doubly* responsible, because we have also put in place the system to get us there.

3.3 Every graduate

The new "business as usual" needs commitment across the board. This means we expect sustainable behaviour from everyone, because everyone has both a vested interest in a sustainable future and must act if we are to achieve it. The most famous metaphor in sustainability is that of Buckminster Fuller (1969). He compared the Earth to a spaceship, which has a limited set of resources and which cannot be resupplied except for energy from the mothership, Sun. Most importantly for us, he saw us all as astronauts, or, more specifically, crew, because "there are no passengers on Spaceship Earth" (Fuller, 1965, as cited in Vallero, 2005, p. 367).

This inclusiveness—the stress on everybody being involved—translates to the professionals whom we train in our institutions. Cortese (1999, p. 6) states:

> Unless higher education responds quickly to ensure that all of their graduates, regardless of their fields of study, are environmentally literate, then it is unlikely that our future leaders will demonstrate the analytical thinking, the will or the compassion to adequately address complex issues such as population, climate change and social equity.
>
> Environmental specialists alone will not help us move toward a sustainable path. A compartmentalized approach further reinforces the assumption that environmental protection should be left to environmental professionals. All humans consume resources, occupy ecosystems and produce waste. We need all professionals to carry out their lives and activities in a manner that is environmentally sound and sustainable.

3.4 The sustainable practitioner

The concept of the sustainable practitioner is fundamental to the Otago Polytechnic approach. It involves applying the lens of sustainability to a person's work. A practitioner in this context is anyone who is working, and it applies as much to those in IT as it does to midwives, lawyers, geographers, mechanics or wildlife rangers.

In order to work through what it means to be a sustainable practitioner in any discipline, it is necessary to work through the skills, values and behaviours that are expected of that profession or trade. One way to do this is to develop a statement starting with these words: "A sustainable practitioner in [discipline] is someone who …"

This should be accompanied by a short narrative that describes the desired behaviours. The focus here is not on specific practice, but on answering the question, "Imagine a sustainable future (in 10, 20,

> **BOX 3**
>
> Imagine a forestry worker, Peter, attending an entry-level-skills chainsaw maintenance course.
>
> As part of that course the future chainsaw operators are taught all about being careful when changing the chainsaw oil, about not spilling it and collecting it for recycling. These are skills that can be assessed upon graduation and the graduates are certified as "sustainable".
>
> However, what is more important is what our graduate does on the first day of work when, after a morning of carefully changing oil, they are instructed to "just chuck it in the stream, you're holding up the whole gang".
>
> And what do we expect our graduate to do when on the first day on the job they are told to go and chop down the last kauri (a New Zealand native tree)?
>
> The answer isn't as simple as telling them to say "no", whereupon they'll get fired and someone else will chop it down. Nor is it as simple as saying "yes", which is clearly unsustainable. Nor is the answer that we should teach integrated catchment management, as such material is probably outside the scope of our chainsaw operator course. Instead the answer needs to involve polite questioning and discussing alternatives.

50 years): what is your discipline's contribution to that sustainable future?" In describing these behaviours we must try to go beyond the trivial, the things that every worker should do (recycling office paper, walking up stairs, etc.), and get to the harder questions about the nature of the profession.

In Box 3 is an exercise to help identify the skills, behaviours and values of any discipline. Although it describes a specific role and set of issues, it could be useful to think through the equivalent situation in your own discipline.

Each department in an institution can work with its stakeholders to identify behaviours expected of their own graduates. What are the equivalent technical skills? What are the arguments for not using these skills? What responses would you offer? What is the equivalent in your discipline to the last kauri tree? Would you recognise it? You might not expect to be told to cut down the last kauri tree, but you might get told to cut down the 999th, so what is the equivalent of the 999th kauri tree, and would it matter? What say the leader of the logging gang says we need to be very productive this month: your colleague's child needs expensive medical treatment, so yes, green is good, but not this month?

This scenario was used as part of a survey of incoming students across the entire institution. Incoming computing students were particularly strident that they would follow instructions even if a task was unsustainable. This seems at odds with the characterisation of our students as independent thinkers.

The concept of the sustainable practitioner easily translates into graduate profiles, learning outcomes, objectives and assessments. Sustainability is about context and the big picture (systems thinking, ethics, evaluating change, scientific and creative paradigms) and a few methodologies (e.g., carbon footprinting, as appropriate). At Otago Polytechnic, the participatory process of articulating the sustainable practitioner for each department has meant that very few people have said, "But I'm a [discipliner], this has nothing to do with me." The response has been the opposite, with a surge of sustainability-related teaching and research.

We can offer a bottom line for each discipline. A sustainable practitioner will be successful when they support people and nature in their actions by:

- enabling people to meet their own and future generations' needs in an equitable way
- causing no harm to nature
- consuming resources at a rate at which nature replaces them
- ensuring nature is not subject to materials it cannot process.

3.4.1 Relationships with the professions

The role of the disciplines is fundamental to building an understanding of sustainable practice, along with the appropriate teaching to deliver on that understanding. The relationship between the professional bodies and the educational institutions is therefore of critical importance. The development of a new business as usual in some cases requires a radical upheaval of what is expected within a discipline—and these changes may be more fundamental than adding some recycling.

As an example, consider hospitality. When Otago Polytechnic rewrote the programme document for the National Diploma in Hospitality (Management) (Level 5), it included an explicit statement describing a sustainable graduate profile:

> Graduates will have an awareness of sustainability issues in the hospitality industry and will be able to apply principles in practice. Sustainability will be integrated into the delivery of the programmes and will be modelled directly for students by the behaviour and attitude of teaching staff. Thus teaching staff must use resources responsibly in the classroom and in their personal work. (Otago Polytechnic, 2008, p. 9)

The statement goes on to identify several specific areas of sustainability, including:

- encouraging use of the most efficient and productive methods (e.g., reducing power outputs, using seasonal products, composting waste and reducing washable linen usage)
- encouraging use of local products where available, including coffee roasted in New Zealand
- demonstrating a commitment to using environmentally friendly products
- increasing the provision of materials to students online rather than in hard copy
- encouraging the construction of professional networks and support structures
- encouraging ownership and responsibility by helping students to realise that social sustainability is the result of everyone's actions, and that each of us must consider the impact we are having.

This last outcome is significant. In hospitality, until recently the response would have been, "In the kitchen you do what you are told", and the best we could hope for was that graduates would hold on to that knowledge for a few years until they were in a supervisory position. Now, even within the acknowledged hierarchical structures of the kitchen, there is recognition of personal responsibility.

Our aim is for the sustainable practitioner to have sustainability issues at the forefront of their mind, which they use as a framework for balancing the demands put on them. This positions sustainability as a framework—not as a competing demand. How does this differ from normal professionalism? A profession is more than membership of a society; it has implicit obligations to a commitment to competence, integrity and morality, altruism and the promotion of the public good within their domain (Cruess & Cruess, 2008; Evetts, 2003; Rice & Duncan, 2006). These commitments, Cruess, Cruess and Johnston (1997) argue, form the basis of a social contract between a profession and society.

3.4.2 The role of disciplines

There are two contrasting schools of thought regarding the role of disciplines. Some see disciplines as the problem: Van Dam-Mieras (2006, p. 15) refers to a "fragmented reality" and Lautensach (2004, p. 12) criticises the "intransigent barriers between disciplines and departments". Lautensach argues that such parochialism leads to a blinkered view, giving a "misleading assessment of risks and novel concepts, diverting our attention from the crisis" (2004, p. 12), and that it reinforces the curricular compartmentalisation of environmental education:

> As long as universities insist on structuring teaching and research by means of the traditional disciplinary boundaries, they will not be able to give full attention to the most urgent problems of the day ... By ignoring the limits of their discipline and getting away with it,

specialists and professional organisations are implicitly, and perhaps unwittingly, communicating to the learner that such behaviour is morally acceptable despite its adverse outcomes. (p. 12)

Charpentier (1994), on the other hand, promotes the role of the disciplines in EfS. Although she advocates a campus-wide initiative, she argues that a programme of EfS should be flexible enough to adapt to each department's knowledge area and curricular focus. Stephens et al. (2008) also see value in the disciplines, but only with the addition of interdisciplinary approaches. Only when the "fiefdoms" allow interdisciplinarity will graduates be able to cope with complex, real-world problems that cannot to be addressed adequately by a single discipline or profession.

Alex Koutsouris (2009) argues for education and research to cross traditional disciplinary boundaries, and describes a continuum of collaboration from multidisciplinarity, through interdisciplinarity, to transdisciplinarity. Multidisciplinarity entails each discipline working in a self-contained manner but working on a common problem; interdisciplinarity is a mixing of disciplines that can lead to new questions and methodologies; and transdisciplinarity gives rise to an "overarching paradigm" that brings together divergent world views, "thus creating new boundaries for exploration and understanding" (Koutsouris, 2009, p. 17).

We have, then, a strong call for a transdisciplinary, sustainable approach to higher education. The sustainability journey is described as a "wicked problem" (Morris & Martin, 2009) because it involves complexity, uncertainty, multiple stakeholders and perspectives, competing values, lack of end points and ambiguous terminology. It means dealing with a mess that is different from the problems for which our current tools and disciplines were designed. For this, we need multidisciplinary thinking, we need transdisciplinary thinking, and we need new thinking.

The driver for EfS within each discipline is a sustainability statement for the discipline. In order to develop this, the stakeholders need to work together to produce a vision for their discipline.

3.4.3 From discipline statements to curriculum

The development of such a vision statement for a discipline can then be used to drive the development of a discipline-based sustainability curriculum, using a three-dimensional model: a cube. The cube has these axes:
- levels on the New Zealand Qualifications Authority framework
- sustainability curriculum descriptors
- discipline and subdiscipline areas.

Each axis has a number of divisions, creating rows and columns of cells. For each cell in this cube, sustainability can be considered in terms of:
- value change
- articulation
- strategies
- legitimacy
- informed decisions
- learning outcome statements
- exemplar teaching strategies and resources.

It is not anticipated that every cell will be filled. Rather, the model provides a "sustainability matrix" whereby a discipline can decide which sections of the cube to fill in.

3.4.4 An example from computing

The breadth of the impact of computing is what drives many of us to take sustainable computing beyond our own footprint. This is recognised in statements such as National Advisory Committee on Computing Qualifications (NACCQ) policy, which makes sustainability a priority for computing education:

Computing and IT underpins every sector of society as a pervasive and influential discipline with global impact. The NACCQ vision is that our graduates, our practitioners and our academics understand the concepts of social, environmental and economic sustainability in order for them to evaluate, question and discuss their role in the world and to enable them to make changes where and when appropriate. (National Advisory Committee on Computing Qualifications, 2007, p. 5)

Several Otago Polytechnic graduates have held titles along the lines of "Editor: Web edition" for various newspapers. Others are in TV, online news sites and so on. Crucially, there has been a significant change recently: these people are making decisions about the news, prioritising and presenting; they are not just technicians for someone else. Emily Nussbaum (2009, p. 1) describes the emergence of a Web team at the *New York Times*:

> The proposal was to create a newsroom: a group of developers-slash-journalists, or journalists-slash-developers, who would work on long-term, medium-term, short-term journalism—everything from elections to NFL penalties to kind of the stuff you see in the Word Train. This team would cut across all the desks, providing a corrective to the maddening old system, in which each innovation required months for permissions and design. The new system elevated coders into full-fledged members of the *Times*—deputized to collaborate with reporters and editors, not merely to serve their needs.

We have, then, a new discipline, one that combines computing and journalism to produce a hybrid with significant impact.

In this new role of developers/journalists, the potential impact is enormous. As the future of journalism becomes programming and interaction with data, and more of our graduates find themselves in this position, the question must be asked, "Are we preparing them for this responsibility?" My point is that in this new era we are going to have to work carefully to ensure the capabilities delivered by the new toys are used appropriately. I doubt that many in computing education could say that we are properly preparing graduates for roles in a new discipline such as this. It serves to highlight for me the need for our focus on sustainable practitioners.

CHAPTER FOUR

Transformation: Process and curriculum together

Sterling (2004a) describes three levels of institutional response to sustainability. First there are the *accommodative* responses—education about sustainability. In these responses, sustainability-related content and/or skills are accommodated into the existing system, often by "bolting-on" modules about sustainability. Second-order responses are the *reformative* responses, which require deeper questioning and reform of the institution's purpose, policy and practice, in order that learning for change takes place. At the highest level, Sterling describes a *transformative* response—education as sustainability. In this transformative response, a living, inquiry-based curriculum is developed. Phillips (2009) holds that:

> It is only through this type of response that education can provide an environment where learners can transcend the limited set of skills offered by traditional education and gain the skills they will need to contribute to a more sustainable society. (p. 2)

The transformation process applies to both the learner and the institution. At Otago Polytechnic we are committed to a mission of inspiring capability. As a rubric, we try to avoid strong distinctions between teaching, research and operations. The mix of all three is where we see successful transformation occurring.

4.1 Separate course or integration?

There is no single template for EfS. Some institutions will choose to add it on to existing programmes; others will opt for a more revolutionary approach. Given a choice between a separate course and integration, I believe sustainability should be integrated into the context of the programme rather than be a separate EfS101: Introduction to Sustainability. This avoids the great danger of students not making the connection between sustainability and the context of their discipline. The worst case scenario is students, having completed a course on sustainability, saying, "Thank goodness that's over, now can we get back to the real work?" In a 2006 study, Carrithers and Peterson (2006) found that such an educational disconnection damages both sustainability and the discipline (business), as "neither faculty group is telling the whole story" (p. 373). This not only leaves students unable to connect the two discussions, but it also confuses and frustrates them in their attempts to do so.

Perhaps it shouldn't come down to a choice between a separate course and integration into the context of the programme. Holmberg and Samuelsson (2006) tackle this question and their answer is: both are needed. They argue that a separate course is needed to give the basic understanding of sustainable development, but that sustainability must be integrated throughout the programme to give discipline-specific tools and conceptual models for dealing with dynamic and complex systems, and to achieve a feeling of how things are interconnected.

4.1.1 A core competency

A different perspective is to consider sustainability as a core competency rather than as a subject matter in its own right. In this way, sustainability becomes a lens (Blevis, 2006) through which we view other things, and which functions in the same way as reading allows us to access other material. Paten, Palousis, Hargroves and Smith (2005) use this approach to describe sustainability as a "critical literacy" for engineers.

Mellalieu (2009) also makes the case for sustainability as one of a range of core competencies that are the foundation of long-term national wealth creation. Mellalieu's argument is that EfS is the academic literacy for the 21st century.

Following initial discussion in 2006, in March 2007 the Otago Polytechnic Academic Board agreed to develop "core capabilities", a process they described like this:

> This will involve core competencies that would be expected of all graduates such as, an ability to communicate effectively, and specific competencies or application of competencies within discipline contexts. Existing programmes will over time through the usual programme review processes progress the development of graduate profiles to encompass the core competencies. (Otago Polytechnic, 2009d, p. 1)

Table 1 sets out the generic capabilities approved by the Academic Board in April 2007 (Otago Polytechnic, 2009d, p.1).

Table 1 Generic capabilities	
Literacy	Is able to listen and read with understanding and is able to communicate effectively both verbally and in writing.
Numeracy	Is able to use mathematical and numerical knowledge to meet the demands of study and work.
Creative thinking	Is able to analyse, evaluate and make informed judgements in study and work practice. Is able to think creatively in relation to study and work practice.
Problem solving	Is able to identify and analyse problems and develop solutions to them.

Information access	Is able to research, access and analyse information from a variety of sources (including current information sources, repositories and modes).
Ethically and socially responsible	Has an awareness of ethical standards and responsible practice which apply to their industry or profession and can demonstrate the importance of working within them.
Autonomous learning	Is able to develop as a learner and take responsibility for own learning.
Operating in teams	Contributes to and functions effectively within work teams, leading by example.
Safe practices	Is able to demonstrate safe working practices and operates safely within the working environment.
Sustainable practice	Has an awareness of sustainability issues that incorporates sustainable practice.
Treaty of Waitangi	Has a level of understanding as an individual and collectively under the requirements of the MoU [Memorandum of Understanding] with Kā Papatipu Rūnaka ki Araiteuru.
Personal effectiveness	Leading others—is able to lead small teams of others. Future focused—understands the need for skills that equip the individual for future developments. Adaptable to change—is aware of current developments and adapts and responds to change. Goal setting—is able to set personal goals. Time management—self-manages time to meet goals.

The challenge was to include the capabilities into courses and programmes in a way that gives guidance to lecturers and provides evidence without unduly affecting the nature of the programme (except where transformation of the programme is the desired outcome).

Table 2 sets out the options for that integration. The options range from whole-course descriptors, through adding new learning outcomes, to specific core capabilities being taught and assessed as part of a course. The preference is to modify existing learning outcomes to reflect core capabilities. This approach recognises that the core capabilities are an integral part of learning in a discipline. It also aligns with drivers to reduce the number of learning outcomes in course documents. This integration has to be a balance of being generic, such that we do not have to list every core capability every time, while being specific enough to give guidance to the lecturer that a particular capability should be included in a specific case.

Table 2 Options for integrating core capabilities

Alignment	Whole-course descriptor	Include new specific learning outcomes	Modify existing learning outcomes	Generic core capability learning outcome	Indicative content/ assessment	Graduate profile, aim, programme outcome
Course is about the core capability (e.g., Sustainability 101)	☺	☺	☺	☺	☺	☺
Core capability provides structure for course (e.g., cradle to cradle in design, triple bottom line in business)		☺	☺	☺	☺	☺
Core capability provides context for course (and vice versa) (e.g., social justice as context for communication teaching)			☺	☺	☺	☺
Core capability provides examples in course (e.g., case study for assignment)				☺	☺	☺
Core capability not relevant for course						☺

4.1.2 Teaching sustainability

The strongest argument against explicitly teaching sustainability is the problem of the raised eyebrow. To some lecturers, sustainability is a fad, a marketing-management inanity that will go away if they ignore it long enough. This is despite, in some cases, using approaches and teaching material that clearly align with the curriculum statements in this chapter. If they felt they were being coerced into teaching sustainability, they might do so with a raised eyebrow, or perhaps a smirk that could undermine the purpose.

The sustainable practitioner approach at Otago Polytechnic has been to encourage departments to develop their own understanding of what it means to be a sustainable practitioner and how this is defined. For some—Social Services, for example—this has not meant including the word *sustainability*. Instead, they have focused on concepts of social justice. In other departments, the discipline's relationship with sustainability is not straightforward. For example, in some areas lecturers saw themselves as teaching best practice, and mismatches between best practice and sustainability led to robust discussion and decisions to make sustainability explicit.

One reason for making sustainability explicit is for transferability. Students in engineering subjects may be taught about energy management, but they need to know that the same principles of conservation apply to setting a road through fragile landscapes.

Two other issues many encounter when making sustainability explicit are the definition and the notion of dogma (see Part 2). I have acknowledged my preference for a pluralistic approach to definitions of sustainability and accept that this does not sit nicely with expectations for a simple, concise definition. Related to this is what Wals and Jickling (2002) call the "shallow consensus", whereby the different values and interests embedded in discussions of sustainability are glossed over. They argue that the concept needs to be openly challenged, negotiated and discussed (see Part 2).

4.2 Sustainability curricula

The focus of this book is on the process of developing sustainability practitioners rather than providing a set of teaching materials for sustainability. This is a significant challenge for some disciplines and is not something that can be prescribed in a sustainability curriculum. However, having acknowledged these difficulties, there is a need for some form of curriculum statements.

The following statement formed the basis of the curriculum approach at Otago Polytechnic:

> As a society we have to learn to live in a complex world of interdependent systems with high uncertainties and multiple legitimate interests. These complex and evolving systems require a new way of thinking about risk, uncertainty, ambiguity and ignorance (Stagl, 2007). These systems require that we can think simultaneously of drivers and impacts of our actions across scales and barriers of space, time, culture, species and disciplinary boundaries. (Mann, 2008)

Taking this statement as a guide, sustainability should not be seen as an extra subject and should not be confused with green issues, or with education about the environment. Instead, it provides a context for learning within and about the student's discipline. If we translate it into skills, it means that our graduates need:

- skills in systems thinking
- an understanding of the connected nature of our socio-ecological system
- skills in critical and creative thinking
- an ability to act as change agents
- an understanding of ethics
- a sense of participation and action
- skills with the toolkit for their specific discipline (if they are an accountant, then we would expect triple-bottom-line accounting, etc.)
- an understanding of sustainability itself (definitions, etc.)

- to have undergone a transformation on the basis of personal reflection
- to hold appropriate values.

Many of these skill sets align with general sets of skills and educational initiatives. These include an emphasis on effective communication, multiple intelligences, systems thinking and student-led learning, along with generic sustainability competencies, such as being able to work in integrated teams to solve problems, lateral thinking and being able to engage with creative ideas outside practice norms.

There is an increasing number of sustainability curriculum statements from disciplines and subdisciplines. Unfortunately, many of these are too narrow to be of much use here. Learning for sustainability aims to go beyond individual behaviour change and seeks to engage and empower people to implement systemic changes. Two curricula that are worth commenting on are from The Earth Charter initiative (Earth Charter International, 2009b) and Second Nature (Second Nature, n.d.).

The Earth Charter curriculum (Earth Charter International, 2009a) takes the same approach as Otago Polytechnic in that it proposes that learning for sustainability should be embedded in the whole curriculum, not included as a separate subject. It is values driven, and the assumed norms—the shared values and principles underpinning sustainable development—are made explicit so that they can be examined, debated, tested and applied. Critical thinking and problem solving are prescribed in order to build confidence in addressing the dilemmas and challenges of sustainable development. Participation is addressed as both a target and a vehicle for involving learners in decisions about how they learn, and in the application of the learning experiences to everyday life on both a local and a global scale.

The Second Nature curriculum statement for sustainability includes the concepts of scale and connectivity. Scale is considered in terms of time, in the sense of both the immediate and intergenerational effects of human activity, and space, in the sense of both the local and global effects of human impact. The human connections between the physical and natural world are also fundamental, underpinned by the simple statement "Humans are a part of nature". Also important are the effects of the physical (built) environment and the natural environment on human health. This latter relationship is expressed in terms of population, consumption, technology and carrying capacity.

Second Nature also proposes paying attention to ethics and values, which cover issues of equity, justice, culture and sustainable development, as part of their curriculum. Students are introduced to different ways of measuring societal wellbeing, and ecosystems are viewed as communities with hierarchies of relationships. The Second Nature curriculum statement also reflects the action approach, with a section on motivating sustainable behaviour.

4.3 Systems and systemic thinking

Sustainability requires a systems approach. People need to be aware that their actions will have impacts, which will be both positive and negative, intended and unintended. They need to understand forms of relationships such as hierarchies, partnerships and feedback, and that humans form part of a complex web. Systemic thinking emphasises patterns, trends and feedback loops. The goal of systemic thinking is to engage learners to recognise and explore connections within and between systems.

Given that we have decided to keep the disciplines, it is worth noting that many authors consider the notion of the subject discipline contrary to systemic thinking. David Orr (1992), for example, gave a goal of ecological literacy as being to foster the quality of mind that seeks out connections. He saw this as running counter to the overspecialisation and narrowness of discipline-based education.

Fox (2009, p. 304) explains this using the metaphor of driving a Mini: "It takes you many complex decisions and movements to keep it on the road but for many individuals it soon becomes easy. Then imagine the fourteen or so people who can apparently fit into a Mini, all trying to do the same task, with chaotic (and comical effect)." Similarly, she says, our professional disciplines restrict us from having a common intellectual framework on a theme.

Almost every author in EfS stresses the importance of systems thinking. Svanström, Lozano-García and Rowe (2008), for example, argue for systemic thinking as a means to cope with complexity and to find balance between different dimensions. Davies (2009) argues that a learning society must be able to think in terms of systems, focusing on "understanding the interactions between human and ecological systems, and restructuring human systems to be more sustainable" (p. 220).

Systems thinking can be contrasted with analytical thinking, but they aren't opposites. Analytical, or reductionist, thinking is about breaking things apart, synthetic thinking is about putting things together and systemic thinking is about combining the two in order to discern the patterns in a larger system, to understand cause and effect chains, and, ultimately, to create changes within and across systems (Svanström et al., 2008).

In a slightly different interpretation, Morris and Martin (2009, p. 156) contend that "learners cannot deal with the wicked problems of sustainability without learning to think and act systemically". Other than in a few disciplines—software engineering, geography and others—systems thinking is not currently the default mode. Strachan (2009, p. 84) characterises the experience of most learners as:

> Moving from a multi-disciplinary approach in their early years, grounded in their limited experience of the world, through to an increasingly reductionist experience in which they become more specialised and less prepared for the inter-connected complexity of the world in which they have to live and work.

While Strachan recognises this is a gross generalisation, it does help to explain how our otherwise impressive disciplines can have such a tendency towards unsustainable outcomes.

This is a real challenge. So much of learning is necessarily reductionist—students need to know the detail of their field—so we cannot just "go wide". As Stephens et al. (2008, p. 321) argue, we need a curriculum that requires this "new and emerging set of skills, perhaps most importantly skills requiring synthesis, integration, and appreciation of complex systems", but as well as, not instead of, a traditional knowledge base.

What are we actually looking for in systemic thinking? Senge, Laur, Schley and Smith (2006) argue that we need more than just the specific tools of systems thinking—dialogue, working with mental models, personal mastery and building shared visions. We also need a broader set of attitudes, values and practices embodied in each one of them. So, how do we go about teaching systems thinking? Remembering that EfS is a learning process, not a product, the question becomes: How do we include systems thinking in our teaching and learning?

Strachan (2009) suggests that just as we need to understand unsustainability in order to fully grasp sustainability, so by highlighting an obvious lack of systems thinking it is possible to demonstrate to learners the nature of systems thinking in a practical way. Morris and Martin (2009) believe that the answer may lie in the difference between a difficulty and a mess. Difficulties are problems that usually have a well-defined and clear boundary, involving few participants, short timescales and clear priorities, with limited wider implications; messes have no well-defined problem or solution, timescales may be long and at best we can only seek to improve the situation.

This distinction between a difficulty and a mess can be reflected in our teaching, and certainly in our assessment. We strive to provide clear, unambiguous problems for students to solve—usually with an expectation of a correct answer. But in providing students with the solutions, perhaps we are giving the wrong message: that perfect solutions are indeed possible. In being more accepting of messes we might be better preparing students for the real world.

Strachan (2009) provides a simple exercise in which learners can apply systemic thinking and discover how everything is linked. Strachan suggests asking learners to select an object with which they are very familiar and investigate it by asking a series of questions, writing the answers on large sheets of paper. The questions can be very simple, such as, "What is it made of?", "Where has it come from?" or "Who made it?" Or they can be more searching, such as, "What needs does it fulfil?", "Is it necessary?" and "What will happen to it in the future?" The task is then for learners to identify connections between their answers, producing a web-like diagram.

If the goal of systemic thinking is to develop thinking about our socio-ecological system, then a second goal for our EfS curriculum must be to actually apply these connections to the connected nature of our socio-ecological system. In short, we need some ecosystem science. How much ecosystem science is needed? Is the focus on people, or physics or plants? The answers are simple, yet not simple. It is not sensible to attempt a list of what people need to know—energy balance, the water cycle, etc. Instead, the answer must be defined by the context. Fortunately, we have made considerable progress here, and the answers are embodied in the notion of the sustainable practitioner.

Take, for example, the role of procurement within computing. The kinds of dilemmas highlighted in the forestry scenario of Box 3 are played out in computing purchasing on a regular basis. Every year organisations purchase hundreds, if not thousands, of computers. How is one of our graduates expected to behave when told to "get them off the back of a truck this year", or to choose between several competing suppliers, all touting apparently green credentials. One of the things we expect our sustainable practitioner to be able to do is to recognise if something is unsustainable. This has two aspects: they need to recognise and deal with "greenwash", and they need to understand the implications of the potential purchase in terms of systemic thinking.

An exercise for students in understanding the connections involved in computing procurement is illustrated in Figure 4, which shows the sources of the components of a typical laptop. The challenge for students is to try to find a component that isn't linked to unsustainable practices, such as environmental degradation, human rights injustice, war, pollution and so on.

Figure 4 Sourcemap representation of the components of a typical laptop

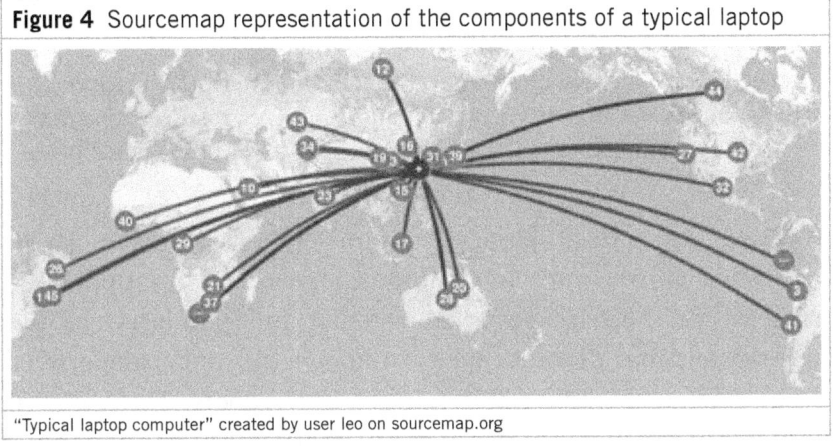

"Typical laptop computer" created by user leo on sourcemap.org

The implication here is not that we should stop buying computers, nor any of the other goods we could similarly explore. It is not possible to buy a "clean" computer, despite whatever carefully managed marketing we might receive. The learning here for students is about the connected system we live in. An exercise such as this puts this information directly into context. (I don't have an easy answer as to what should be done with this information. Should we avoid such tainted computers? Yes. Is that possible? No—but we can buy fewer, perhaps having a four- instead of a three-year replacement cycle.)

I propose an *inclusive* consideration of ecosystem, which is one that makes little distinction between natural and human systems.

In recognising the oneness of systems, ecological literacy becomes a broad concept. It therefore embraces cultural literacy (Polistina, 2009). Indeed, many authors argue that the cultural aspects are pre-eminent: "Culture includes our whole system of beliefs, values, attitudes, customs, institutions and social relations. The global crisis facing humanity is a reflection of this system and is therefore a cultural crisis" (Polistina, 2009, p. 121). Polistina (2009) describes cultural literacy as a fundamental skill that learners need in order to develop sustainability literacy. She argues that:

> reflection on our own culture and other cultural systems can help reveal the complex social, environmental and economic relationships that need to be changed to make a successful shift towards sustainability. (p. 121)

Paralleling cultural literacy is interdisciplinary literacy. Again, given that we have made a commitment to the disciplines, we have to find mechanisms to ensure that our graduates are able to bring together different types of knowledge and perspectives. If not themselves, being able to work with others. Tormey et al. (2009) describe how the very expertise that is necessary to address sustainability problems can often get in the way of finding solutions (see also Stephens et al., 2008). Students therefore would benefit from experiencing frameworks of disciplines working effectively together. Tormey et al.'s advice is to reframe one's own discipline as a culture, rather than the right or "normal" way of doing things.

In this context, Sterling (2009, p. 79) draws a distinction between systems thinking and ecological thought:

because while ecological thinking is systemic (relational), systems thinking is not necessarily ecological. Systems thinking can be used as a methodology for antiecological, as well as ecological, ends. Yet at the same time, systemic thinking can help sow the seeds of an ecological worldview, it can help facilitate the critical reflexivity—or deep questioning of assumptions.

Here my interest is on the reflexivity—in these terms: self-reflection. Why would we want someone to do that? The answer is simple, we want them to care. To achieve sustainability, "First you have to care" argues AtKisson (2008, p. 15). The systems thinking we require must be connected to a system of ethics and values.

4.4 Critical and creative thinking

> The study of environmental problems is an exercise in despair unless it is regarded as only a preface to the study, design, and implementation of solutions. (Orr, 1992, p. 94)

Critical and creative thinking are essential for learning approaches to EfS. EfS challenges us to examine the way we interpret the world and how our knowledge and opinions are shaped by those around us. Critical thinking involves purposeful reflective judgement concerning what to believe or what to do—it is a systematic process of inquiry, learning and thinking. Creative thinking involves looking at problems or situations from a fresh perspective that may lead to unexpected, unorthodox solutions.

For Senge et al. (2006), critical and creative thinking are paramount to sustainability. They describe a process whereby the first question is to unearth all the assumptions underlying strategies for the future, then to ask how current strategies serve us if these assumptions change, and finally what options we can create to improve outcomes in the event that the assumptions change.

Martin (2005) describes the "backcasting" approach, which contrasts with the usual way of approaching the future through forecasting. While forecasting starts from where we are and projects forward, backcasting works like this:

> The idea is to think imaginatively about the business or organisation to which you belong and seek to explore a range of fundamental changes that will make it more closely fit the sustainability framework. From each alternative future created, you then work your way backwards from the future towards the present in stages, asking such questions as—what barriers did we overcome; who helped us; who did we need to persuade? (p. 168)

This process engages people in conceiving and capturing a vision of their ideal future, and in uncovering the steps required to get there. It clearly relies on both critical and creative thinking.

4.5 Ethics

For Fagan (2009, p. 1), the ethical imperative is the basis of sustainability:

> To live a particular lifestyle that, knowingly, impacts detrimentally on a neighbour—be that an individual living in the next house—or a country in the next region, cannot, arguably, be tolerated. To know of poverty in the economically developing world and not use that knowledge to act to relieve it, could be considered unethical. This position holds profound implications for politicians, schools and universities.

Ethics is about doing the right thing and about doing good. A sustainability paradigm gives us a framework for doing good, for working towards coexisting on a crowded planet with limited resources where our actions can adversely affect others. As Thomas Berry (2000, 2009) remarked, our moral concerns should include biocide and ecocide, as well as homicide and genocide.

Immanuel Kant's ethical stance was the categorical imperative: act as though your behaviour could become a general rule. That is, ask: Could everyone else do the same? Can your behaviour become the norm? For example, how sustainable is it if everybody were to drive to work in a car with three empty seats?

We explored the sustainable practitioner chainsaw scenario (Box 3) with a group of building trades lecturers. They agreed with the premise that their graduates should act as sustainable practitioners, but felt that this should not extend to challenging behaviours that they considered unsustainable. It is, they say, vital to the safety of the building that you do exactly as you are told on a building site. Fair point, but this is in itself a value position—that of safety. We further explored the ramifications of safety. What would we expect our new graduate to do if instructed to climb on an unsafe structure? They would be expected (required even) to object to this immediate threat. Clearly, in the area of safety the new graduate on the building site is empowered to manage their own safety. The same applies if they see someone else doing something dangerous—they are required to object. So, let's say they are instructed to do something unsustainable—say hide some heavy metal in material destined for landfill, or to order rainforest timber—the timescale of the threat might have changed, but it is still there. We don't have actual answers for what people should do in these situations. Clearly the line is blurred, and sustainability does not possess the same urgency as a wobbly ladder, but the point is that everything has a basis in value positions—and students should be aware of these positions.

4.6 Willingness to act

Perhaps the most significant development in sustainability in recent years is a recognition of the need for action. While this has long been the case on the individual and small scale, we now increasingly recognise that actions speak louder than words. It is critical, then, that our sustainable practitioners are able to make things happen—that they are change agents. AtKisson (2008, p. 34) describes these people as "the sales force". He says:

Change agents help the innovators translate their ideas to the mainstream. They learn how to 'work the system', so that new ideas can begin beneficially infiltrating the cultures, institutions and organizations where they are sorely needed.

Key to the success of a sustainable practitioner approach is that graduates not only know what the right thing is, and how to go about doing it, but that they *want* to take action. Being a change agent involves doing more than making changes for oneself, but exactly what is involved depends, to a large extent, on the discipline. Remember Schendler's "find your biggest lever and use it"? This lever will clearly be different depending on the nature of the job.

So, then, we need all of our graduates to be change agents. What does this entail and how do we get there? This is not an area we can afford to get wrong. As we know from the chainsaw scenario, getting fired for being obstinate will not help either a graduate's career prospects or sustainability. As Robinson (2009, p. 130) argues, we "also need skills in effectively and persuasively presenting the proposed changes, sometimes in difficult circumstances if the change goes against the ingrained culture of the organisation".

There is a wealth of material on what it means to be a change agent. A sample of the skills and attributes required includes:
- research skills (quantitative and qualitative)
- accurate observation, monitoring and recording
- questioning
- action planning
- negotiation skills
- report writing, visual and oral presentation skills
- a customer-service focus
- a systems approach
- learning from every decision
- recognition of a new business as usual.

The change agent role still has the other elements of sustainability—systemic thinking and critical and creative thinking—as part of the core skill set. The sustainable practitioner does not promote change for the sake of change. Rather, he or she takes AtKisson's (2008, p. 9) advice: "Doing sustainability requires, first and foremost, that we stop and think":

> It takes time for people to understand something new, to reorient their thoughts, to imagine what a change or innovation will require of them. Change Agents need to be willing to let an idea ripen in the minds of others. (AtKisson, 2008, p. 242)

I am reminded of the welcome sign at the wildlife sanctuary in Wellington: "A journey that will take 500 years". Clearly there is urgency, but we also need to take the long view.

So, how should we best engage students in learning to be change agents? Again, the pattern is familiar. In education terms, this change agent concept invokes an action competence approach (Jensen & Schnack, 1997). The focus is not necessarily on a course on, say, effective change management (though that could be a good idea too). Rather, the key is in the hidden curriculum.

Lautensach (2004) argued for a "pedagogy of liberation" (p. 16) that "guarantees that the learner will actually act on such (sustainability vision) insights" (p. 5). Lautensach contended that we need to specifically "help empower and motivate the learner towards taking action" (p. 5). The aim is for an institution that is actively "doing the right thing" itself, and providing opportunities for student involvement—not just on the periphery but as active partners.

CHAPTER FIVE

Transformation

The framework for sustainable practice presented in Chapter 1 describes a process for transforming education to encourage the preparation of sustainable practitioners. It involves the collaborative development of the goal of being a sustainable practitioner in each discipline, the articulation of this in curriculum documents for each discipline and the progressive development of learning that aligns with this goal. This chapter explores in more detail the process of getting there.

It is widely argued (e.g., by Placet, Anderson, & Fowler, 2005) that making small improvements while maintaining the status quo is unlikely to result in the required changes for a sustainable future; and further, that the radical innovation needed cannot be achieved by an education system that adopts a "more of the same" approach (Alvarez & Rogers, 2006; Bekessy, Samson, & Clarkson, 2007; Sterling, 2004b; Wals & Jickling, 2002). Sterling (2004b) argues that sustainability should not be seen as another issue to be added to an overcrowded curriculum, but instead should be "a gateway to a different view of curriculum, of pedagogy, of organisational change, of policy and particularly of ethos" (p. 50). He argues that:

> A full response, however, commensurate with the size of the challenge, implies a change of educational paradigm—because sustainability indicates a change of cultural paradigm which is both emergent and imperative. (p. 50)

A transformative education paradigm refers largely to two factors: the transformation of the learner and the transformation of the education process to achieve that.

5.1 Journey, not destination

In explaining the commitment to sustainability at Otago Polytechnic, I often use the line: "This is a journey and we might never reach the destination." That the transition towards sustainability is an ongoing responsibility has implications for the way we manage the change (Stephens et al., 2008). Most change management process literature describes a one-off change—a single implementation, or a transition to a changed state—and while it might take a while, there is the clear implication that you'll get there eventually. However, sustainability is not an achieved/not achieved binary state: the goal might never be reached.

I like the work of Steve Benford (Benford, Giannachi, Koleva, & Rodden, 2009) and think his ideas are particularly applicable to EfS, both in terms of the learning experience itself and the transformation of an institution. Instead of one-off activities, Benford considers interactions to be a continuity of experiences or journeys. This leads to the idea that interactions can be interwoven, steered and that each of us has multiple continuous trajectories:

> While these journeys may pass through different places, times, roles and interfaces …, they maintain an overall sense of coherence; of being part of a connected whole. These journeys are steered by the participants, but are also shaped by narratives that are embedded into

spatial, temporal and performative structures by authors. They are also influenced by the dynamic process of orchestration ... Finally, they may be undertaken by groups and/or involve encounters among participants. (Benford et al., 2009, p. 712)

Sterling (2004a) places an emphasis on capacity building and action in his classification of institutional responses: learning about change/learning for change/learning is change. At the highest level, or transformative response, he describes a living inquiry-based curriculum. Sterling goes on to describe the risks associated with such a transformation; in particular, that we are not certain what a sustainable society will look like, which makes it difficult to know precisely what or how to teach to achieve it. However, he says that all individuals, whether leaders or not, have to learn to be adaptable and take the risks that will help society create a map to sustainability, because the risks of doing nothing are immense. I see a bigger risk, which is that the transformation Sterling desires is so dramatic that few academics, if any, can live up to this specification.

Milne, Kearins and Walton (2006) are critical of the use of the metaphor of "sustainability as a journey" in the professional and business literature. They found that the journey metaphor is commonly applied to both commitments and actions (behaviours, decisions, etc.) that might be considered to lead towards sustainability, and to the process of reporting on the triple bottom line. The idea of journeying, Milne et al., conclude:

> offers paradox and complexity on the one hand (as a potential excuse for relative inactivity and lack of substantive progress) while also expressing a notion of progress, if not actual achievement even in the embarkation on the journey itself. (2006, p. 813)

The concept of a journey, they suggest, and its associated imagery in organisation and management writing, appears not dissimilar to Alice's aimless wanderings.

With Milne's warning ringing in our ears, my argument here should not be interpreted as avoiding the hard stuff. Indeed, in several discussions early in the journey at Otago Polytechnic I made a case against focusing first on programmes that had obvious sustainability angles (outdoor education, horticulture, etc.) and instead argued that we ought to demonstrate transformation in accounting, computing and hospitality. I was worried that focusing on "low-hanging fruit" would risk complacency; and worse, that others in disciplines with a longer journey to travel might abdicate responsibility by assuming that sustainability belongs in courses that are already about the environment. In practice we approached this problem from both ends: getting early results with the low-hanging fruit and working out strategies for the high-hanging fruit at the same time.

5.2 Top down or bottom up?

Otago Polytechnic has benefited from committed and enthusiastic senior management as well as seeing many initiatives develop from the grassroots—from groups and individuals, staff and students. Pam Williams (2008) has studied the organisational communication structures within universities with a view to understanding how to encourage the development of EfS. She found that both hierarchical, transformative, connected leadership and distributed leadership are required. Further, these two forms of leadership can be enhanced by strengthening the networks between them. In other words, the development of EfS should be both top down and bottom up, simultaneous and integrated.

Herrmann (2007, p. 77) also argues that the development of EfS is "neither a top-down nor a bottom-up issue but is a shared responsibility for each individual member of that institution". Others have found that top-down leadership has failed to materialise and argue that proponents of sustainability need to develop strategies that do not assume a top-down approach, but warn that relying on a ground-up approach may not achieve the cultural shifts that are a precondition for mainstream sustainability (Bekessy et al., 2007). Phillips (2009), on the other hand, advocates a ground-up development and says that administrators should learn to become comfortable with transformation stemming from all areas of the institution.

5.3 Telling the story

An important aspect of the transformation of the institution is the internal storytelling. To this end, Otago Polytechnic has a marketing representative on the EfS management group who has helped the group to develop some principles to underpin sustainability communication: to "tell the story". They are:
- make sustainability front of house—a new "business as usual"
- celebrate success, but don't make claims to have achieved things we haven't
- remember that it is a journey, not a destination—the message is "this is the path we are on"—and make no claims to be perfect
- be open to debating issues of sustainability
- tell real stories, even ones where not everything has gone right
- where possible, personalise stories
- treat everything we do in the organisation as a learning opportunity.

Tools for telling the story have included the sustainability newsletter, signage around the campus (especially interpretive, work-in-progress signs and operational signs for things such as recycling, see Figure 5), CEO updates, staff awards and champions.

Figure 5 Living Campus signage

My Otago Polytechnic colleague Katie Ellwood and I also produced a short book, *A Simple Pledge: Towards Sustainable Practice* (Mann & Ellwood, 2009), which was formally launched on World Environment Day 2009. Although there were a number of reasons for producing the book, its main purpose is to convey the message of the institution's commitment to staff and students. As part of the required culture change, we need to continually reinforce the message that this is the journey Otago Polytechnic is on, that you are part of it and that you can do it.

The book, which is quite short, was designed by two final-year communication design students, Craig Scott and Simon Horner, and is deliberately optimistic in tone and style. The book uses the right-hand pages to tell the story in words and pictures (Figure 6) and the facing pages to provide background on each aspect of the story. The articles cover the range of teaching, community outreach and operations, and the images present students, graduates and staff working towards being sustainable practitioners in a range of ways. These images are overlain with diagrams of sustainable thinking to give the message that everything can be viewed through a lens of sustainability.

Figure 6 Sample pages from *A Simple Pledge*

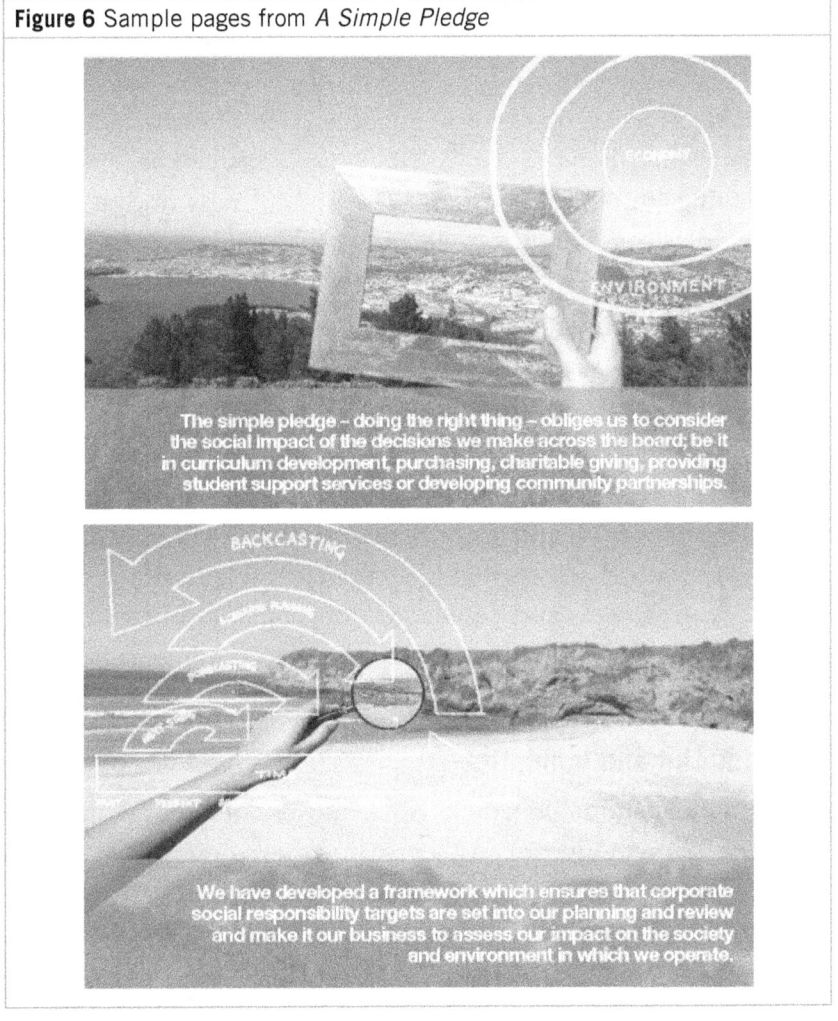

We also developed a pledge card to prompt people to take ownership of the messages of the book and, hopefully, to make a commitment. More than 200 staff have signed this public pledge:

> As a member of the Otago Polytechnic community, I pledge to become an ambassador for sustainable practice.

5.4 A new role for educators

One of the changes that is required in a transformation to EfS is the acceptance of a new role for educators. In the area of sustainable practice, academics cannot hope to portray themselves as experts (Stephens et al., 2008). The concept of teachers co-constructing knowledge with students is not new and is one of the principles stemming from the influential 1977 UNESCO Tbilisi Intergovernmental Conference and enshrined in other sets of principles (Eisen & Barlett, 2006):

> Help faculty explore the shift in pedagogy from a paradigm of teacher as expert to teacher as facilitator of learning, becoming co learners with students and with each other. (p. 26)

Freire and Freire (2000) note that an education that is all about imparting answers is inadequate. Howard Gardner (1993, of multiple intelligences fame) argues that it is important to educate people so that they can ask new and interesting questions. Within the context of the discipline, the EfS educator becomes a facilitator and enabler of change rather than a disseminator of knowledge (Polistina, 2009).

5.5 Footprint and handprint

Acting as a sustainable practitioner means both reducing one's footprint (reducing harm) and increasing one's handprint (actions towards sustainability[2]). There are other ways of referring to the same thing: McDonough and Braungart (2002) talk of "doing more, not

---

[2] Handprint: Action Towards Sustainability is an initiative of the Indian Centre for Environment Education (http://www.handsforchange.org/)

just less", while Senge, Smith, Schley, Laur and Kruschwitz (2008) refer to "regenerative sustainability" (p. 38).

Lautensach (2004) extends these principles to education in terms of the moral duty to mitigate degradation through education. For those of us working towards being, and encouraging others to be, sustainable practitioners, this means that:
- as educators we must address our impacts *and* engage in actions that promote sustainability
- the biggest lever we have in promoting sustainability is to prepare our students for careers as sustainable practitioners
- our graduates also need to address both their footprint and handprint, and the skills and values they acquire should reflect both these aspects.

The need for a dual focus on footprint and handprint can be clearly demonstrated in computing, which we have already used as an example in relation to sustainability in procurement practices (see Section 4.3). Computers have transformed the world in which we live—they have become commonplace at work and at home. There are currently more than one billion computers on the planet (Global Action Plan, 2007), which, through manufacture and disposal, have a very large impact. The Global Action Plan reports that between 2000 and 2006, energy consumption from nondomestic ICT equipment rose by 70 percent, and that in 2006 ICT equipment accounted for 10 percent of the UK's energy consumption. James and Hopkinson (2009, p. 24) observe that "Few people realise that many of the gleaming devices on their desktop, or in specialist facilities, are effectively coal-fired, with all the wastage and pollution that implies."

This tells us that graduates in computing, particularly those with a hardware focus, should have skills that include procurement, electronic waste minimisation, energy management (including server virtualisation and consolidation), storage management and operational structures (e.g., contracting for service rather than servers).

> **BOX 4**
>
> In a recent year at Otago Polytechnic, a department printed a whopping 670 pages per student. Another department just topped this with 671 pages per student. These numbers are too high, and those departments are working to reduce their printing. But here's the awful truth: those two departments print far less than anyone else!
>
> In total we print more than 6,000,000 pages per year. Four and a half million of these are in the academic departments for an **average of 1,387 pages per student**. This number is way too high in financial and environmental terms. Five departments top a mind-boggling 1,900 sheets per student, the most being 2,048 pages per student (the numbers here are expressed as b/w page equivalents). Our annual paper use would make a stack 662 metres tall. That's more than twice the height of Auckland's Sky Tower and only just shorter than Mt Cargill.
>
> Knowing this, we're working towards a goal of zero printing to students. We accept that some things do require paper—a complex diagram on a worksheet, for example. But we are even more sure that other things can't be justified: one-page-per-slide handouts of PowerPoints; class lots of whole chapter printouts, etc. What messages does this high volume send to our students? What technical things could we do to reduce this printing? How do we prepare our students for this initiative?

The ubiquitous and enabling nature of computing, however, means that the opportunities for doing good—extending the handprint—are even greater. The 2008 round of the Computerworld Excellence Awards (presented by Computerworld New Zealand) for the first time included an award for sustainable ICT. All of the finalists achieved tangible and sizeable sustainability outcomes. For most, these were more sustainable ways of running their business operation, primarily data consolidation and server virtualisation projects. However, Airways New Zealand, the winner of the Sustainable ICT award, went much further than this in using technology in an innovative way to boost sustainability.

Airways New Zealand has demonstrated the potential of ICT to dramatically affect sustainability far beyond its own footprint. The CAM (Collaborative Arrivals Manager) reduces aircraft fuel emissions by reducing waiting time while airborne. This initiative uses an interactive Web application that manipulates demand for take-off and arrival. Now, the new online system allows airlines to manipulate landing slots themselves, using drag-and-drop functions on a website. The system helps airlines pick their most important flights, such as those with international connections. It also means that delays now happen on the ground, not in the air, thus saving considerable amounts of fuel. Interestingly, and perhaps significantly, this same project also won the independently judged award for Excellence in the Use of ICT in Customer Service.

Having looked at these isolated examples from IT and computing, in Chapter 6 I take an in-depth look at the issues relating to creating sustainable practitioners in this discipline. I've chosen IT for this in-depth examination both because it is my area of expertise and because it is an interesting example to explore, as the link to sustainability may not be immediately obvious.

CHAPTER SIX

Sustainability in IT and computing

6.1 Incoming student survey

In 2008, 539 new students at Otago Polytechnic responded to a survey designed to probe their attitudes towards sustainability (Shephard, Mann, Smith, & Deaker, 2009). It was undertaken in the first weeks of the first semester, so the students had not had much exposure to teaching at a tertiary level, but they had chosen their fields of learning. The survey asked students about sustainability values, activities and relevance (the findings are reported in full in Mann, Smith, Shephard, Smith, & Deaker, 2009). Here, the focus is on the responses of the students in information technology (IT). Demographically, the 60 IT students were not significantly different from the general intake, but IT does have a strongly male bias, with 83 percent of the students being male, whereas the institution as a whole is only 38 percent male.

Some departments are strongly pro-ecological, especially the health departments. IT students have a slightly pro-ecological world view but are strongly anthropocentric in some areas, believing that the balance of nature is strong enough to cope with impacts, that humans will work out how to control nature, that the ecological crisis has been exaggerated and that humans are meant to rule over nature. The gender effect is significant, but it is not the whole story: business students (70 percent female) are not significantly different in their world view from IT students.

Respondents were asked to rank a set of potential issues. Priorities for IT students are largely the same as for others, with protecting environment having the highest priority. Beyond that, IT students are more concerned about strengthening the economy and limiting population growth, and less concerned about creating a fairer society, than other students. Although IT respondents were familiar with popular concepts such as renewable resources, the greenhouse effect, etc., they were less confident with specific ecological concepts, such as carrying capacity and the precautionary principle. Incoming students from across the institution reported limited understanding of indigenous concepts of sustainability.

Although the majority of IT students saw the relevance of sustainability to their study, those who did not were quite outspoken about the suggestion. Other disciplines had similar proportions who doubted the relevance of sustainability, but none showed the strength of feeling of IT students. This is a comment from an IT student who thought sustainability was not relevant to the discipline:

> IT is about humans and technology, not humans and the environment.
>
> We are studying IT! The complete opposite of nature.
>
> We only use paper and computers.

Another student saw a little more relevance: "It is something worth considering but not relevant enough to warrant drastic action." Some were motivated by self-interest ("It'll put food on my table in the future"), while others saw the need to reduce IT's impact ("IT consumes a lot of natural resources").

Those who saw high relevance also had a range of views. Some described computing's impact ("computing has a high carbon footprint"), while others could see that the resources needed by the industry might run out. Some looked past these immediate considerations to a broader view:

> No matter what field we go into, if the environment is disrupted enough to cause flooding, e.g., sea level rise, and greenhouse gases to grow, then it is important to recognise the implications.

> ... because everything in this world is relevant to sustainability I regard it as high in my chosen field as if there were no sustainability, I would have no career.

There are very interesting patterns here. IT students have anthropocentric world views and see a low relevance for sustainability, and business students, despite being mostly female, share these characteristics. The health fields have people with the strongest pro-ecological world views but don't make the connection between this and their careers. The anthropocentric engineers, meanwhile, see sustainability as strongly relevant to their careers. The patterns are complex and not well understood, but, clearly, different pedagogical strategies would be needed for these different groups.

There is much variety in the extent to which people undertake, or are prepared to undertake, sustainability-related actions. Surprisingly few students would report unsustainable activity, and IT students showed a reluctance to encourage someone else to change an activity or practice that was harmful to the environment. This is of concern as we try to move beyond our own footprint to using computing as an enabler for a wider sustainability. As I have already noted (Section 3.4), incoming IT students were strongly of the view that they would follow instructions, even if a task was unsustainable. Reassuringly, 50 percent would at least talk about it.

Many authors have identified gender issues in computing and computing education (Bair & Cohoon, 2004), and gender differences in environmental attitudes have been extensively researched and reviewed (e.g., Zelezny & Schultz, 2000). Research since 2000 suggests that women report more environmental concern than men and greater participation in pro-environmental activity. The introduction of sustainability concepts into IT curricula may help to address the traditional gender imbalance in IT intakes by making the IT world more appealing to women.

The crucial sentence in computing education for sustainability is "our goal is that every graduate may think and act as a sustainable practitioner". Being a sustainable practitioner means more than having technical skills: we have to come to terms with world views, affective learning and action competence. This research at Otago Polytechnic has made a start towards understanding these factors as they relate to computing students.

6.2 Curriculum initiatives

6.2.1 Posters

As part of an introductory computing course in the Bachelor of Information Technology at Otago Polytechnic, my colleagues Dr Karen Love and Darrell Love asked students to research a "green" IT poster:

> I wanted the students to research the issues in the area of green computing, but I also wanted them to see that there were things they could do personally and professionally. I decided they needed to present the environmental problems, consequences, possible solutions, and methods for achieving those solutions. Quite a big ask for a single poster. (Love & Love, 2008, p. 266)

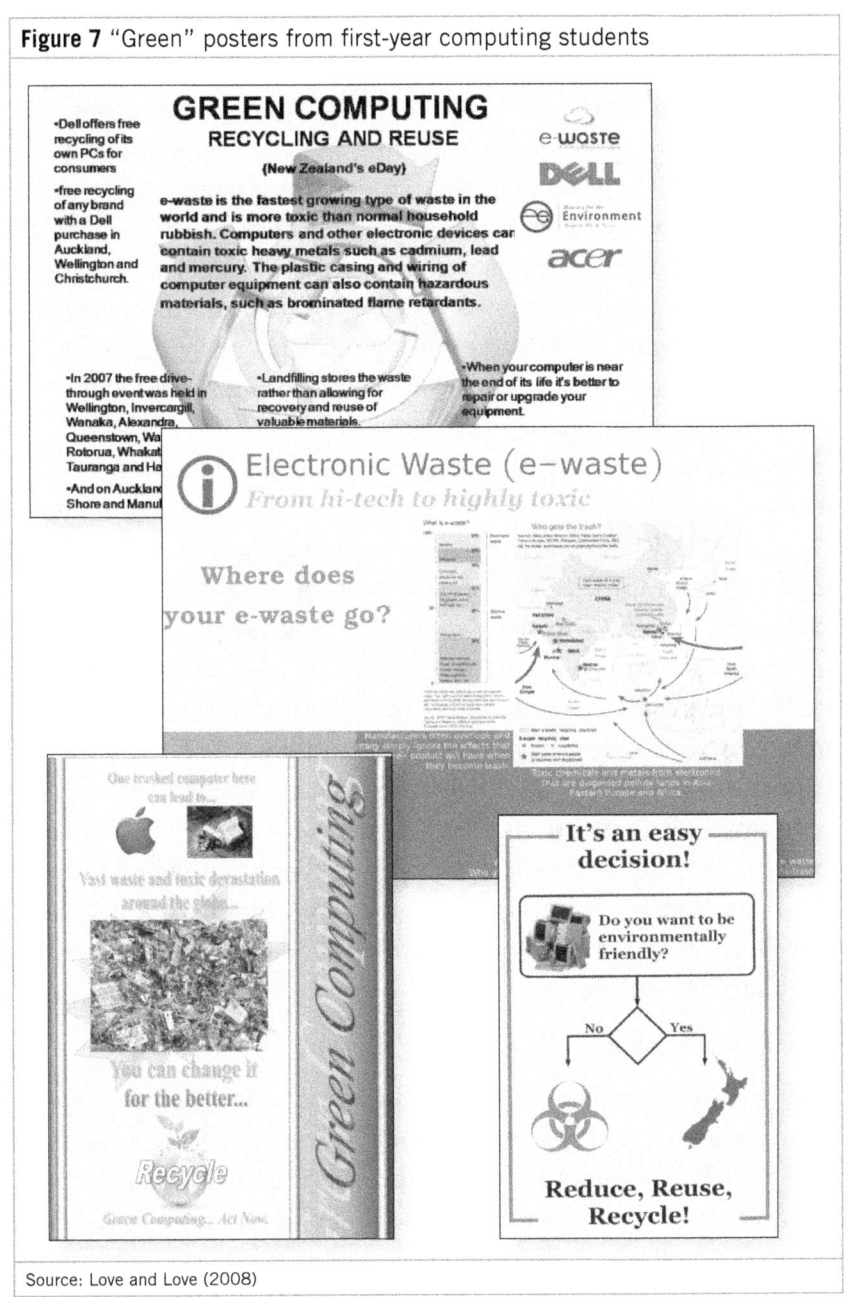

Figure 7 "Green" posters from first-year computing students

Source: Love and Love (2008)

On completion of the assignment (see Figure 7 for a selection of posters), the Loves surveyed the 20 students who had participated. The students were asked to begin by describing the poster assignment in terms of its effectiveness in raising their awareness of "green computing". Nineteen out of the 20 students said that the assignment had been effective or very effective in raising their awareness of the need to approach their industry from a green perspective, and most said that they would not have learnt what they did on their own. Whether or not this will lead to behaviour change is unknown, but the students themselves believe that what they have learnt will cause them to make some changes in their own behaviours.

The students revealed a surprising grasp of the conflict between the needs of a profit-driven economic model and those of environmental sustainability. While they think that awareness of an environmental need will prompt action, they strongly believe that it will take economic incentives to truly motivate the general population to use computers in a more environmentally responsible fashion. The students focused almost entirely on reducing the environmental impact of computing manufacture and disposal. Their primary response to what the individual who uses a computer can do was dominated by two actions: "turn it off" and "recycle it". A few students showed greater insight, adding that manufacturers need to take responsibility for the sustainability of their products, and four students noted that the Government needs to be involved in stimulating recycling.

6.2.2 Communication

Another assignment required students to view a film or read a novel from selected lists, and then to interpret it in order to identify issues deriving from the interrelationship between human beings, technology/IT and the environment (Love & Love, 2008). The students were asked to identify the solutions proposed by the material, to discuss the consequences of inaction and to recommend actions individuals could take to promote sustainability. This was the rationale for the assignment:

> I responded with an assignment which I believed would challenge my BIT [Bachelor of Information Technology] students to 'see' the cultural dialogue surrounding technology, the environment and human survival by engaging with image and narrative in the form of novels or films. Why? First, it is in our communication, both through images and narratives, that beliefs, attitudes, behaviours and values are expressed, critiqued, and passed on. Our social, political, and cultural ideologies are both transferred and challenged by our stories and our images. Second, it is because books and films are 'hot' media—that is, they are participatory media requiring active engagement in order for the receiver to intake their content; thus, they often elicit a more profound response than 'cool' or passive media. (Love & Love, 2008, p. 267)

Love and Love (2008) report that the students demonstrated surprising insight in their interpretations. Here are some examples of their comments:

- *Jurassic Park*—"An unsustainably simple structure bluntly forced upon a complex system." Questions raised included: Should we do it? Do we understand the risks? Is profit the main driver? Can we stop after we've started? Are the benefits worth it?
- *The Island*—"Can we build a sustainable business and maintain a sustainable planet?" Issues discussed included bio-fuel or human food supply as a problem of human worth, and new technologies cannot be allowed into common use before ethical issues have been explored.
- *Fight Club*—"Do nothing so nothing is done." Issues canvassed included not using or developing technologies is a valid option; suicide rates rise in parallel to technological and economic development; and economic sustainability—simplifying the way we live by liberating ourselves from consumerism and debt.

- *Johnny Mnemonic*—"All my life I have been comfortable in my own corner, looking out for number 1—now, I'm responsible for the whole world." Messages included the shining city vs. the rubbish heap—realising our dreams is destroying our reality, and reconsidering what we think is necessary to our lives.
- *An Inconvenient Truth*—"Technology empowers human life and holds the key to our future." Technology and environmental sustainability must be developed together.

Several students argued that technology must be developed along with a full assessment of its potential impact on the environment. After completing this exercise, these computing students demonstrated surprising insight into sustainability.

6.2.3 Software engineering

In the second year of the Bachelor of Information Technology, the software engineering course is based on the strategy of making it real—real projects for real clients—and takes an empowerment approach (Mann & Buissink-Smith, 2000; Robinson, 1994; Smith, Mann, & Buissink-Smith, 2001). The Capstone projects (see Section 6.2.4), for which software engineering is preparatory, follow an integrated methodology that combines elements of both agile and structured software development. This Agile Development Framework (ADF) approach, described more fully in Mann and Smith (2006), focuses on the production of robust working systems (software, hardware and maintenance documentation) rather than on planning, development documentation and processes, which are important but are simply means to an end.

In 2008 the software engineering classes developed the information systems that power the Living Campus. They worked with the Living Campus team to identify their needs and to produce systems to meet those needs (see Figure 8). A rewarding outcome of this was

SUSTAINABILITY IN IT AND COMPUTING

the extent to which the realisation that the Living Campus is more than just a garden came through in the students' work.

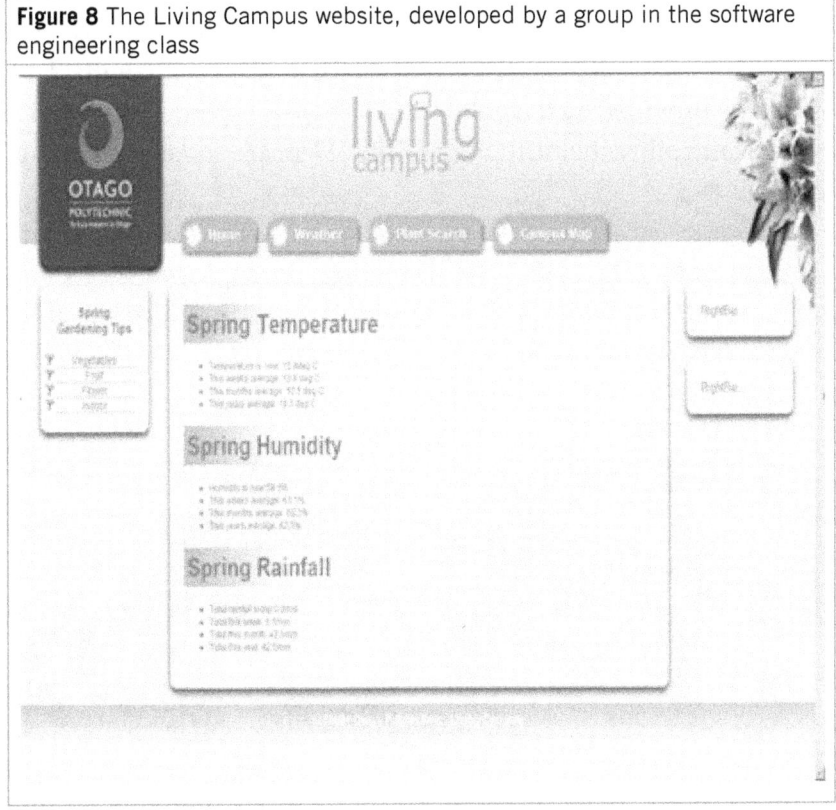

Figure 8 The Living Campus website, developed by a group in the software engineering class

Their systems not only supported the functional aspects of operating a garden, but also enabled the process to be used to enhance learning. Groups within each of the two classes worked together to produce integrated systems with a common database, style guide and directory structure, and the groups worked to pass information from one section to another—from maps to search, and back again.

6.2.4 Capstone projects

A feature of the Bachelor of Information Technology at Otago Polytechnic is the Capstone project. Students work in small groups in their third year to produce a working system for an external client. Students are responsible for finding clients to work with, with some assistance provided by the polytechnic. There is a strong interactivity and sustainability theme to the projects. Here are some examples of the projects:

- CityScape—an immersive panorama that will allow visitors to explore any city where the technology is deployed (currently Dunedin) and express their creativity, building a community narrative (see Figure 9).

Figure 9 CityScape

- a GPS-based virtual walking tour of central Dunedin's historic landmarks
- Ant Wall—an interactive virtual ant colony display to attract more visitors into a redeveloped nature area of Otago Museum (which achieved the 2009 NACCQ Award for student research)
- a carbon emissions management system for agriculture
- an e-training system for the staff at Dunedin Casino, to facilitate training in identification and management of potential problem gamblers

- AidTraining Aid—a design for a Web-based training system to prepare volunteers to be ready for volunteer work in other parts of the world
- PestWeb—a system that will store interactive information about pests and pest management to allow farmers easy access to information
- a dynamic dust monitoring system for rural roads
- Windmill—an integrated system for monitoring and managing the client's innovative power-generating windmill
- MarineQuest—interactive learning material for the Portobello Aquarium
- Eduvaka—literacy and numeracy training for Pacific Island preschool education
- a farming system modelling game
- a virtual catchment model and community website for Kaikorai Stream.

6.3 SimPā

6.3.1 Overview

It is critical that sustainability not be consigned to being a one-dimension problem—in computing terms, solely as the need to reduce consumption of energy. Social and cultural elements are not just part of sustainability—they are fundamental to it. This is illustrated by the SimPā project, which is a multi-year collaborative project designed to strengthen and communicate Māori culture, tikanga and knowledge using innovative approaches.

SimPā, first described by Mann, Russell, Camp, Crook and Wikaira (2006), was a collaborative partnership between the Otago Polytechnic and Kā Papatipu Rūnaka ki Araiteuru. In short, the project aimed to provide a means of telling Māori stories in three-dimensional (3D) game format.

The project recognises that Māori culture is a vital part of what distinguishes New Zealand from the rest of the world. It is intended that the project will help in the creation of 3D game-based digital content so that distinctly Māori voices, stories and cultural content can be encouraged and promoted. The importance of Māori digital content is key to the Government's Digital Strategy:

> Māori are both creators and consumers of content and distinctively Māori content is particularly visible in the areas of: broadcasting; the arts and creative industries; as well as the education, health, and business sectors including tourism. Māori digital content is important not simply for its economic potential, but also as a vital means of expressing Māori culture in today's society and into the future, strengthening Māori society and identity, telling Māori stories to other Māori, and communicating with the wider world. Hence the importance of content being created and maintained in the Māori language. (New Zealand Government, 2005, p. 12)

The SimPā project was funded in 2006 by the Digital Strategy through the Department of Internal Affairs. The project was completed in May 2009. The original objectives are presented here, with a view to exploring how this project can be seen to contribute to a wider view of sustainability.

The SimPā project has demonstrated to us the value of focusing on the handprint aspects of being a sustainable practitioner in computing. We have helped the rūnanga retell their stories to themselves. In their new form they are still retained within each rūnanga. The participatory methods have seen the development of digital interactive storytelling with Māori culture, tikanga and knowledge. The SimPā project uses gaming software to create various "GamePā", which are virtual environments based on actual places. At the start of the project the team proposed a process of participatory development for each rūnanga. For each group they saw a process of helping the community identify important stories and then converting these stories to a game-based environment. The project has several objectives, which are described below.

6.3.2 A participatory approach to game development

Here the objective is to bring people from the papatipu rūnanga together to learn about their own place, share their stories and convert this knowledge to digital form. This objective has been achieved, with members of the rūnanga engaged in the process of developing digital Māori content through a series of wānanga.

6.3.3 A software tool for creating specific Māori virtual environments

The "SimPā toolkit", including a suite of software packages and game templates, was created. However, looking back, the toolkit has been more about a process of engagement, partnerships and ideas than about the technology. Although the project promised game-based narratives, the team spent much time engaged in discussing the wider applications of digital technology. For instance, one rūnanga has a long-held role as archivists for the iwi and they see the potential for SimPā to help with this role.

6.3.4 Use of GamePā to teach Māori concepts

Although the stories are hosted in a game environment, this was used as a platform for further engagement, such as the recording and production of documentary-style interviews that share stories of the past and present. These resources give access to the mātauranga they convey as well as the knowledge used to create them. They also provide a means of access for members, regardless of their location in the world.

6.3.5 Techniques and practices for the use of GamePā

We did not expect the rūnanga to move so quickly to using the skills learnt through the SimPā developments to develop further applications. The best thing to come from the SimPā project was the initiatives beyond the original concept. The project has extended beyond traditional stories to include contemporary narratives, such as the story of Puketaraki's new carvings and Moeraki's expedition to Te Papa.

6.3.6 A new specialist area in education: Māori digital content

We believe that a new subject area is emerging as a result of the partnership between rūnanga members and the technology team at Otago Polytechnic. We see constantly evolving partnership as a very positive outcome and believe that this model of engagement could be used for further partnerships.

6.3.7 Spreading this initiative beyond the collaborating partners

This collaboration involves complex structures of knowledge ownership, an important part of which is to maintain the integrity of specialist knowledge and tikanga.

CHAPTER SEVEN

Sustainability in other disciplines

In May 2008 each head of department contributed to a half-day session designed to capture and share the progress of each discipline towards EfS. Presented here is a sample of reports from departments outlining where they are on their EfS journey. Departments were asked to reflect and comment on three things:
1. What does it mean to be a sustainable practitioner in your field?
2. How is this being reflected in your programmes (graduate profiles, learning outcomes, etc.)?
3. Evidence of how this is making a difference in teaching and learning.

7.1 Art

The visual arts have a major role to play in how communities operate, and the wider contextualisation of art in the programme ensures that students understand the ways in which their work both cements and critiques the social order.

On a more practical level, the pragmatic aspects of healthy and sustainable art practices have led to the gradual movement from toxic to nontoxic materials in all subject areas, and the development of an understanding of the correct use and disposal of chemicals and other waste materials. Some artworks are designed to endure for a long time, and consideration is given to the ways in which they are made and conserved. Other artworks are temporary, and consideration must then be given to the question of the disposal and reuse of the material used in their construction. The developing field of digital art has also led to an increased awareness of how communities are created and maintained in the digital environment, and the sustainability of the digital infrastructure.

Graduates have an understanding of the principles of sustainability. They are able to evaluate their work in relation to its socioeconomic contexts and the ways in which it supports the social fabric, and can recognise strategies for mitigating environmental and social harm in the conceptualisation and creation of their artworks and in their practice as a whole. The graduate profile includes an understanding of the role artists play in sustaining the cultural and spiritual life of the community, and of the pragmatic elements of the philosophy of sustainability in terms of the care and conservation of resources, and health and safety.

The School of Art at Otago Polytechnic runs a regular programme of public seminars, lectures and workshops, within which the principles of sustainability find a direct focus. Issues of sustainability are integrated into supervision, feedback, recommended reference materials and the theoretical and practical framing of research projects.

All staff members in the School of Art are aware and proactive concerning sustainability issues. Contemporary artists and arts enquirers are arguably more attentive to these issues than most other groups of teachers in tertiary environments. Reasons for this vary, from the critical role of the artist in contemporary society today, to the need to protect oneself and students from toxic substances in the studio. The current focus on sustainability at Otago Polytechnic provides an opportunity to scope the field of focus within the School of Art programmes. Through this process it has become clear that this focus encompasses a wide range of concerns, from everyday vigilance on a practical level to a sociopolitical critique within teaching, learning and research.

It is crucial that students understand the context in which the need for the notion of sustainability has developed, and the cultural, social and emotional implications of the current situation. Without an understanding of the impact of these ideas on people's ability to hope and plan for the future, or to respect the past, the incremental steps needed may seem too difficult, or may meet with an emotional resistance that is hard to understand. For these reasons, it is important that students gain an understanding of the histories of the development of global capitalism and the impact on the planet of the apparently sound notion of an increasingly higher standard of living.

Some knowledge (for example, of the complex mixture of the desire for a universal higher standard of living and the strategy of planned obsolescence) is necessary. It is also important for people to understand and deal with the emotional implications of the situation; for instance, the sense of grief experienced during a period of species and habitat loss. The concept of nature is very complex, but its oversimplification is in part responsible for the situation we are in. These issues can be made explicit in much of the art theory and history syllabus, in areas of Otago Polytechnic where emotions and their expression are foregrounded.

7.2 Design

The Department of Design teaches degrees in product, fashion, interiors and communication, and also has significant numbers of students in foundation courses. The Department of Design has signed up to The Designers Accord,[3] which is "a coalition of designers, educators, researchers, engineers, business consultants, and corporations working together to create positive environmental and social impact" (The Designers Accord, n.d.,a). The vision of the Accord is to "engage all members of the creative community in a dialogue about the importance of integrating the principles of sustainability in all practice and production" (The Designers Accord, n.d.,a).

The Accord makes some strong statements about the role of the designer, including being proactive: "all adopters to engage in conversation about social and environmental impact with every client and customer, and integrate sustainable alternatives in their work" (The Designers Accord, n.d.,b). The Designers Accord and the earlier Design Manifesto (Barnbrook, 2007; Holland, 2001) make clear the wider responsibilities of designers in looking at moving beyond the designer as a transparent communicator and instead recognising an ethical responsibility for the messages portrayed.

The Department of Design has adopted sustainability as a core concept in its teaching. For example, in the Master of Product Design degree, sustainability is seen as a core principle when developing new products.

In other design programmes, EfS is integrated across courses and forms the basis of both specific courses and elective options:

> Sustainable design principles are more explicit in all specialties to encourage understanding of the social, environmental and economic impact of design decisions and allow students to embed sustainable design in their practice. Students develop an understanding of their own culture in order to better consider the social and cultural practices and perspectives of others. (Otago Polytechnic, 2009b, p. 1)

---

3 http://www.designersaccord.org/join-us/

In addition:

> In 2008, staff worked together to develop a statement about what sustainable practice meant in a design context: Sustainable practice in design is complex and contested. Multiple frameworks emphasise different values. The role of the designer is to explicitly address this complexity in ways that meet both current needs and the anticipated needs of the future. (Otago Polytechnic, 2009b, p. 2)

The School of Design is integrating sustainability through all programmes at course level through inclusion in learning outcomes and assessments and as major project themes. Examples of such projects can be seen in the Bachelor of Design (Communication) which used the Sustainable Campus as its client for the 2009 Year 3 course Interaction Design. In 2008, third-year interiors students designed a sustainable hall of residence in Interior Design Studio 5 and had to research and apply Green Star rating systems, source sustainable materials, implement low-energy principles and consider how to encourage residents to act more sustainably.

7.3 Midwifery

The underlying philosophy of midwifery is aligned with sustainability, as midwifery promotes low resource use and minimising unnecessary intervention. Midwifery practice is about community-based primary health, strengthening family relationships and health promotion, so to some extent the rise of sustainability can be seen as affirming the approach midwifery has always taken (Davies, 2007).

The school has taken the decision to integrate EfS, starting with a new foundation course, BMSD107: Sustainable Development. After this, sustainability is integrated into all second-year courses, with specific objectives for each. In practice courses there is a focus on advocating practices such as breastfeeding, health maintenance, education of whānau, birthing in primary maternity facilities or home birth, etc. In the third year, sustainability is applied to midwifery practice through a Sustainable Midwifery Practice course, which picks up sustainable practice in terms of managing a small business, sustaining oneself in practice and sustainable practices in midwifery practice. The challenge is to promote sustainability and midwifery as a primary health model in a context where medicalisation dominates the maternity services.

Further changes will see a new programme design and delivery model to make midwifery education sustainable for both students and midwifery staff. It is about keeping students in their home locations, and utilising local maternity facilities and resources as much as possible. Working in collaboration with the Christchurch Polytechnic Institute of Technology (CPIT) ensures best use of resources, which are shared.

7.4 Nursing

The School of Nursing argues that health is intrinsically about the sustainability of communities. The school works closely with the Nursing Council, where the descriptions of professional responsibility include:

> competencies that relate to professional, legal and ethical responsibilities and cultural safety. These include being able to demonstrate knowledge and judgement and being accountable for own actions and decisions, while promoting an environment that maximises client safety, independence, quality of life and health. (Nursing Council of New Zealand, 2009, p. 4)

The school believes EfS is already in the graduate profile and learning outcomes are embedded in its holistic approach. Many course outcomes include statements like this one:

> The aim of this course is to prepare health professionals to practise in an holistic context. With reference to anthropological and sociological concepts it is intended that students will develop an understanding of the richness of human diversity and develop attitudes commensurate with culturally safe practice. (Otago Polytechnic, 2009c, p. 37)

In practice areas the school has examined its processes. They have realised that, for example, when teaching how to put on sterile gloves, the practice gloves don't actually have to be sterile.

7.5 Occupational therapy

Occupational therapy as a profession has always maintained a focus on sustainable practice. The profession has its roots in the Arts and Crafts and Moral Treatment movements, with extrinsic activities seen as being the way we communicate our intrinsic beliefs to the world. It is a core belief of occupational therapy that humans need to be involved in meaningful activity, and that we gain connection to the world through activity such as constructing a home or cultivating a garden. Occupational therapists are concerned with manipulating activity contexts (e.g., social, physical, vocational) rather than people. The aim is to meet people's goals, wants or needs through sustained involvement in meaningful activity, allowing people to be as independent as possible. If we change the context a person operates in, it should be for their betterment and independence. As a profession, occupational therapists need to be mindful of the resources and materials they use or put in place to sustain activity, and not burden either services or clients.

During an industry liaison workshop, occupational therapy professionals were asked to describe what a sustainable occupational therapy practitioner would look like. Elements they identified included that they would:

- be advocates of human rights, social justice, occupational justice
- be prepared to accept and/or embrace change
- consider the work/life balance of their clients and themselves
- use leisure and play as primary occupations
- utilise real, not simulated, environments
- value time, both their own and the clients', making every encounter count

- understand professionalism
- ensure equipment is carefully chosen, taking into account client needs, and sustainable practices.

The Occupational Therapy School has set up a Social and Sustainability Committee, comprising both staff and students. The focus of this group is to foster, support and sustain the Occupational Therapy School community for staff and students, as well as to support, introduce and encourage any notions or ideas that encourage sustainable practice within the school.

Notions of sustainability and the management of resources and materials are addressed in a number of stage one courses, notably Adaptive Living Occupation, Humanities for Occupational Therapy and Fieldwork 2. All of these papers have a focus on humans' involvement in activity and what sustains this involvement. A constructivist approach is taken, with students often doing or teaching activities.

In Adaptive Living Occupation, students start to investigate humans' involvement, both present and historical, in activities broadly grouped under the categories of food and horticulture, games, craft and engineering, and design. Students begin with activity workshops, which look at participating in selected activities under these headings (including two weeks with the Horticulture department). At all times there is a focus on doing from scratch, making do with limited resources and not wasting resources or materials. In the second half of this course students then have to plan, resource and present six activity workshops to members of the general public, with the above focus being prevalent.

In Humanities for Occupational Therapy, students investigate, via imaginative literature, film, guest speakers, art and poetry, the commonalities and expressions of human occupations and what sustains them. Notions of humans as makers and sustainers of place are considered, along with the ways we all connect and maintain our place in the world through labour, work and play. Students are asked to present on and involve their class in an activity they partake in, taking into consideration the history of this activity.

In Fieldwork 2, students spend three to four hours a week for 14 weeks with a community in the Dunedin city region. These communities all have a focus on meeting an occupational need for a selected group of people. Students are expected to fit into the community as participants and facilitators of meaningful activity. They are expected to develop an understanding of how activities are managed, resourced and promoted, and how this enables the sustenance of these communities and those who attend.

Other courses include Adaptive Living Technology Stage 1 and Design for the Individual Stage 2, in which students look at the design of equipment and accommodation, focusing on concepts of inclusivity. Students consider how a home or public building is designed in order to limit the number of physical and other barriers, and to explore designs that allow for the greatest flexibility of use. Equipment is evaluated for reuse, modularisation, etc., and students also look at suppliers of equipment and consider their policies for sustainability.

7.6 Sport

Sustainability is inherent in the outdoors ethos, as well as in the systems approach to sport performance. Adventure offers the first step in developing a motivation for sustainable practice through helping students to develop a connection with the natural world. A sustainable practitioner in Adventure would have a local focus and a minimalist approach, yet Adventure is primarily a transport-intensive activity, and sports focuses on human-centred performance, which poses a significant dilemma.

There are many examples of EfS in the Diploma in Outdoor Leadership and Management, including a full second-year paper,

Environmental Science and Education. In this paper students investigate a current environmental issue and present it to the class, calculate the ecological footprint of their own house or another building, keep an eight-week diary on what they are doing to live lighter on the Earth, run environmental education activities for another group of students, deliver an interpretive lesson on a local natural environment, learn about tikanga and te reo Māori and a Māori view on the natural world, and take some action for the environment.

Staff training has been initiated to help staff understand their own actions. The aim is to continue internal discussions on how to instigate sustainable practices throughout our programmes and within the staff and students via individuals, department actions and curriculum.

The school has set a series of goals specific to the school and an action plan to realise these. The goal is for staff to become aware of what sustainable practice means to them personally and professionally, both through sustainability being a permanent agenda item on staff meetings and by building sustainable teaching tasks into course outlines that relate to sustainable practice in social, environmental and economic issues.

7.7 Veterinary nursing

Veterinary nursing practice teaching is based on the use of best-practice standards. Although these are largely health and safety-focused, there are elements of sustainability. The school is taking a paper to the Animal Nursing and Technology Board on the subject of sustainability in relation to all our veterinary nursing programmes and what we are doing to ensure the veterinary industry is producing graduates who are confident about sustainability. The paper recommends including sustainability within the national training standards.

First, sustainability should be looked at in conjunction with health and safety, Treaty of Waitangi, internationalisation, flexibility in teaching and facilitation, financing, resourcing, teaching and curriculum development, and in particular developing all of these areas into the core competencies. Doing all these things responsibly should lead to sustainable practices.

The term "intellectual sustainability" may be new, but the concept is not. For many years a cornerstone of Otago Polytechnic philosophy has been the production of a lifelong learner. The National Diploma in Veterinary Nursing is focused on the production of lifelong learners. Specific behaviours include engendering a commitment to students taking responsibility for their own learning, and developing values such as self-awareness and self-motivation to stay up to date in an everchanging industry.

Sustainability is embedded in all our veterinary nursing programmes, but it needs to be articulated more within the curriculum and through assessment. When we are teaching health and safety and ethics, sustainability is taught without *saying it*—we need to start saying it!

The school continues to work to incorporate new initiatives to ensure the maximisation of sustainability practices into its business plan. All staff within the School of Veterinary Nursing encourage students daily to take responsibility for and perform to best-practice standards, and to become aware of the impact on the environment of other than best-practice standards within the industry. In particular, students are encouraged to think about the choices they make on a daily basis and how these will affect other students, the community and the future.

All staff implement environmental sustainability by using resources carefully and respectfully. Waste is minimised, correct disposal methods are practised to avoid damage or harm to the environment or others and recycling and reuse are encouraged by all staff and students. All course materials have been redeveloped to be compatible with electronic delivery, whether as part of a face-to-face programme or via distance learning.

Specific behaviours include a commitment to innovation. Working effectively for the future means constant adjustment to future needs within the veterinary and animal-related industries as they are developed or identified.

PART TWO

BARRIERS TO INTEGRATING SUSTAINABILITY INTO ACADEMIC PROGRAMMES

CHAPTER EIGHT

An overview of barriers

So far we have looked at a framework for integrating sustainability into academic programmes in a way that engages every learner. This is already happening: sustainability is on the agenda of most institutions. However, in classrooms, in departments and throughout institutions there are discussions going on about sustainability. Sometimes it is pointed questioning, sometimes curiosity, sometimes partial agreement, sometimes belligerence. Here is a selection of common objections that can be heard:

If sustainability is so important how come it doesn't even have a good definition?

Education is the problem, not the solution.

How is this different from the tree hugging we did in the seventies?

Sustainability is just a fad.

Sustainability is still contestable.

Sustainability is too negative, and without vision.

I'd rather the students were critical thinkers.

This is a whole specialist field of study.

Get over it, the world will be alright, it doesn't need saving.

The term *sustainability* has been hijacked and debased to the extent that we should avoid this altogether.

We've tried it before, no impact on graduates.

And so it goes on. In Part 2 of this book I examine some of these objections, or *barriers*, to sustainability and look at how they can be addressed or overcome. For clarity, the barriers have been grouped as follows:

- Is there really a problem, and is sustainability the answer?
- Is it the responsibility of higher education?
- I can't see this working in our institution.
- I want to integrate EfS, but how do I do it?

Subsequent chapters will look at each of these groups of barriers in turn, but first some general comments on identifying and overcoming barriers.

8.1 Identifying and overcoming barriers: The theory

A range of barriers hinder people's ability or willingness to act in a sustainable manner. We need to get better, firstly, at recognising the complexities of these barriers, and then at developing strategies and tools to help people to make successful change.

There are two common approaches to overcoming barriers. The first is compulsion, which in this case would mean a simple requirement that all lecturers integrate EfS. This is certainly part of the toolbox at Otago Polytechnic, where annual staff performance reviews include demonstrating an alignment with institutional strategy. However, compulsion alone will not work. Putting aside (for the moment) issues of academic freedom, it is simply not possible to specify what would be expected in every instance. The goal of sustainability is too complex and understanding of what is involved changes. Also, change by force undermines intrinsic motivation and can backfire as people reassert their own positions and autonomy. How much better to see people enjoying their teaching and responding creatively and critically through choice.

The second approach is to provide information. Information is clearly a requirement for both acting as a sustainable practitioner and preparing others for this path. Information is the basis of intelligent decisions, and fortunately there is a lot of it about, yet still we wander perilously close to disaster. Information is clearly not the entire answer. Sterling (2004a) points out that 30 years ago environmental education was based on the assumption that people were insufficiently aware of the environment and understood little about it. However, Downing, Cox and Fell (2004) found that four out of five people surveyed agreed with the statement "Do you feel reasonably well informed about what you personally could do to be more environmentally friendly?" and they could identify specific actions. Asked which of these environmentally friendly behaviours they actually undertook, only around one in 15 said that they regularly did more than just a handful of these things they had named. Information doesn't always equate to a response or a change in behaviour, and just as I have previously argued that education about the environment is unlikely to result in sustainable practitioners, neither is information about EfS likely to engage lecturers.

There is a third way. Michael Lambert (2004) describes elements of successful change in behaviour. The core elements are:
- a person's readiness
- the environment is conducive to change
- relational factors—positive regard, congruency and empathy—are aligned.

Providing support for behaviour change is clearly not as simple as providing more information. Instead I aim for a third way. Some information is required, but most traction is gained by focusing on empowerment.

Recognising that 100 percent sustainability is a perfect state that is practically unattainable by anybody or any system is an empowering position. It accepts that no matter what their behaviour, everyone can be seen as being on a journey towards sustainability. Too often, as people committed to sustainability, we make the mistake of criticising our fellow travellers, demeaning them as incapable or unwilling to embrace the sustainability paradigm. Too often the assumption is that someone acting unsustainably doesn't know about—or doesn't care about—degradation of the Earth's systems. We make the mistake of then deluging them with facts about the state of the glaciers, or ocean acidity levels, or the perilous state of … I believe we can get much closer to understanding and addressing the real barriers by asking people about the issues *they* face.

8.2 What we know about barriers to EfS

Ferrer-Balas et al. (2008) identify several barriers to transformative EfS within higher education:
- the leadership and vision that promote EfS not being reflected in the assignment of responsibility and rewards
- the bottom-up structure of academic freedom (this barrier can also be seen as an enabler, empowering early-adopter sustainability champions)
- overly rigid incentive structures that do not recognise faculty contributions to sustainable development
- lack of desire to change, especially at the leadership level, reflecting investment in existing teaching and research
- little pressure from society to make changes in the characteristics of graduates and research, giving a university little reason to make transformations.

The Higher Education Academy (2006) examined the discipline-based Subject Centres that have been set up in the UK to provide support for teaching and learning in higher education, and found much variation in progress towards embedding EfS. The Academy went on to identify barriers and suggest actions to overcome them:

- An overcrowded curriculum. Create space through a rigorous review of existing curricula to ascertain what is already there in terms of the development of identified EfS skills and knowledge.
- Perceived irrelevance by academic staff. Develop credible teaching materials which are fully contextualised and relevant to each subject area. This will help ensure that EfS is integral to the curriculum and not a "bolt on" element.
- Limited staff awareness and expertise. Invest significantly in staff development and capacity building through institutional staff development units and Academy Subject Centres.
- Limited institutional drive and commitment. Develop a credible business case for tertiary institutions, setting out triple-bottom-line benefits. Review and amend institutional mission and policy statements.

Velazquez, Munguia and Sanchez (2005) undertook a large literature review of sustainability initiatives in higher education. They found that progress on campuses has not been as fast as expected and identified a range of systemic and individual factors. The problems they describe include lack of a clear and shared understanding of what EfS looks like; insufficient resourcing (both money and time) devoted to it; lack of training in how to do it; and little awareness, interest, involvement or support from administrators, and resistance to change at many levels.

For Holmberg and Samuelsson (2006, p. 8), the barriers to implementing sustainable development in higher education stem from a mismatch between sustainability and the structures of higher education:

> It must be hard to find something more multi- and transdisciplinary than sustainable development. It is also quite clear that the traditional discipline-based structuring of knowledge and research are here to stay. This combination constitutes a major challenge for the universities when implementing learning for sustainable development in higher education.

In overcoming what they describe as the several-fold challenge of university culture, Holmberg and Samuelsson point to the need for integration and separation in both the teaching and organisation of EfS. In teaching they argue that both separate courses and programmes and an integrated perspective throughout the whole education system are required. In terms of organisation, EfS needs to become the responsibility of every discipline, but also of the organisation as a whole.

In Costa Rica, Charpentier (1994) studied the barriers to EfS in the Universidad Nacional system. She found much enthusiasm toward environmental learning, but this was not matched by actual behaviour. The most significant barriers were those stemming from respondents' misgivings about their competence to infuse their courses and curricula with an environmental dimension, and their perception that valued colleagues do not support these actions. She classified barriers according to the source of the obstacle (and the respective characteristics):

- A behavioural belief is one in which an individual associates the inclusion of an EfS in courses and curricula with a favourable or unfavourable consequence. This gives rise to an individual's positive or negative evaluation of including EfS in courses and curricula.
- A normative belief is related to the probability that an important referent (such as a chairperson, colleague or reference group)

would approve or disapprove of the infusion of EfS. This can be described by the subjective norm—the social pressure from important referents and the person's motivation to comply with the referent's desire.
- A control belief considers the availability or unavailability of requisite resources and opportunities, including time, ability and financial resources. This can be real, or could be based in part on past experience, or second-hand information (experiences of colleagues), giving a perceived behavioural control—the individual's judgement of the ease or difficulty of infusing EfS into courses and curricula.

Franz-Balsen and Heinrichs (2007) also highlighted the importance of norm-supporting structures in their study of the University of Lunebur. For people they described as "sustainability distant", an important approach was the identification of the university leaders who shared the mission. They also stress the importance of not being overbearing—respecting the right to not be bothered.

8.3 Structure of barriers and responses

The chapters that follow discuss specific barriers to the integration of EfS. Having identified each barrier, I present some ideas or resources that may help move us forward. An important part of this is to acknowledge the audience's perspective and to provide choices on how to make changes. I also try hard not to make a judgement about the barriers that confront people and institutions.

The goal is that the reader can reflect on whether the barrier is relevant to them, can review their motivation to overcome the barrier and can choose from some things they would consider doing to overcome the barrier. The focus is on small and practical steps, like "Try asking a student . . .", or "In your next lecture try linking your . . . theory to current events." For example, take the barrier "Our vice-chancellor says we need to protect academic freedom." In this case, the perceived barrier is external to the person, who sounds as though they are already motivated to pursue sustainability. What could they do to try to effect change? Actions could be at different levels and might include such things as inviting the vice-chancellor onto a class panel to discuss campus sustainability.

Internal barriers are perhaps harder to address. This is not a textbook in psychology, but understanding the nature and origin of different types of resistance helps us to understand how best they might be overcome. Miller and Rollnick (2002) suggest that categories of resistant behaviour can be reframed and used to create a momentum for change. The types of resistance they identify include arguing, challenging, discounting, hostility, interrupting, talking over, cutting off, negating, blaming, disagreeing, excusing, claiming impunity, minimising, pessimism, reluctance, ignoring, inattention, no response and sidetracking. In all cases, the approach is to explore the barrier through the use of probing questions such as "Do you believe that . . . ?", "Are you saying that . . . ?" and "Does this mean that . . . ?" For example, in response to the view that we have got a while before degradation really matters, we could ask, "Do you believe that we can sustain our current lifestyle and the effect of that on the environment forever?" or "Do you believe that enough individuals can make a difference? How many might that take?"

There is no formula for each barrier. The barriers can be internal or external and of many different types, and the responses have to vary accordingly, as we will see.

CHAPTER NINE

Is there really a problem, and is sustainability the answer?

9.1 Ecosystems are not really under stress and declining, affecting our future

In this first group of barriers the focus is on system degradation and on the people to whom we need to provide the "Why?" of sustainability before we go on to consider the actions of being sustainable practitioners. There are three aspects to this barrier: first, that degradation isn't happening; second, that we're not causing it; and third, that it doesn't affect us. This barrier is perhaps one of the few where more information may change people's views and provide motivation, given that all three of the statements above are demonstrably false. But it is my contention that because few people would report this as a barrier it should not become the focus of EfS.

We clearly need an optimistic approach—I see myself as an optimist. This problem is soluble. But wearing rose-tinted spectacles makes it difficult to see the warning signs. Being overly optimistic could lead one to think the planet's life support systems will keep absorbing human negative impacts indefinitely. Nature is robust and can recover. The dustbowl of the United States recovered, fish stocks can recover (albeit slowly), the ozone hole may be closing, we may stumble across hitherto unknown carbon sinks. However, other activities have recovery times taking many human generations: mercury in fish, radiation half lives (and genetic mutations) and the proportion of lost topsoil that is replaced is probably miniscule. The wilfully ignorant over-optimists (Berry, 2008) project the present and past into the future and assume that business as usual, with some tweaking, is a sensible guiding principle.

Over-optimists can point to the fact that many first-world cities are cleaner than in any time over the last 200 years, and that bacterial and viral assaults on our species have largely been contained. However, in most cases, the avoidance of disaster or chronic hazard has been due to decision making informed by scientific observation. Affluent countries, in particular, have had the ability to legislate against ecologically damaging practices. Education produces the knowledge needed to deal with these challenges. Optimism based on the human capacity to find and implement solutions has a reasonable foundation. Optimism based on "mother nature will take care of it" and the idea that we have a miniscule capacity to affect the biosphere does not.

Unsustainable human activity is, by definition, a real problem. It leads to a situation of ever-decreasing returns, whereby the essential is sacrificed for the comparatively trivial. Unsustainable behaviour cannot continue indefinitely. An unsubstantiated (emotionally driven) position on an issue amounts to pushing your own agenda, whereas the evidence is that many human activities are wasteful and deleterious to life. Not to consider the challenges facing your discipline is to be irresponsible.

The Millennium Ecosystem Assessment (2005) catalogues both the state of the environment and the effect of this on us. Here is an excerpt:

> People everywhere rely on ecosystems and the services they provide. So do businesses. Demand for these services is increasing. However, many of the world's ecosystems are in serious decline, and the continuing supply of critical ecosystem services is now in jeopardy.
>
> An ecosystem is a dynamic complex of plants, animals, microbes, and physical environmental features that interact with one another. Ecosystem services are the benefits that humans obtain from ecosystems, and they are produced by interactions within the ecosystem. Ecosystems like forests, grasslands, mangroves, and urban areas provide different services to society. These include provisioning, regulating, and cultural services that directly affect people. They also include supporting services needed to maintain all other services. Some ecosystem services are local (provision of pollinators), others are regional (flood control or water purification), and still others are global (climate regulation).
>
> Ecosystem services affect human well-being and all its components, including basic material needs such as food and shelter, individual health, security, good social relations, and freedom of choice and action. (p. 3)

The Industrial Designers Society of America (2001) makes the same point—that nature can survive without humanity but that we are dependent on the biosphere for crucial services. Society's systematic destruction of the biosphere threatens nature's health and its capacity to sustain human society.

9.2 If we do need to act, we've got a while yet

It is technically correct to say that we are a long way from being able to actually destroy the world. If by "world" you mean a planet with a biosphere that supports some kinds of life, then yes, you're quite right. However, a world in which humanity can thrive is under imminent threat. The question then becomes, "Do we want to leave such a degraded land for future generations?" So if by "world" you mean the relatively salubrious conditions that humanity has thrived in over the last 10,000 years, your optimism may be unwarranted. To declare that everything will be all right given the plethora of evidence of our species' despoliation of the Earth's ecosystems is to gamble with a stake that doesn't belong only to us. Mother Nature is the house and the house always wins. If you think business as usual is an adequate response, you're not paying enough attention.

Even if a degraded planet is acceptable, on the way to this degraded state things will get very unpleasant. The 1987 World Commission on Environment and Development led by Gro Brundtland identified a number of global issues that might fall within the broad area of sustainability, including:

- the burden of debt in the developing world, inequitable commercial regulations and a growing number of the world's population living at or below subsistence level
- overuse of nonrenewable resources and growing competition for limited water supplies, threatening armed conflict over access to water and mineral reserves
- reduction of biodiversity and increasing desertification
- pollution of air, water and soil, with detrimental influences on the global environment and climate change
- continuing growth of the world's population, coupled with additional economic pressures caused by increased life expectancy
- increasing nationalistic, political and religious extremism, terrorism, armed conflict, mass migration and social disruption.

Many commentators convey a sense of real urgency:

Realism is important because our problems are more urgent than ever. We don't have time to fool around or to be fooled by the delusion that we're making progress if we're not. We don't have time for theories that aren't grounded in the real world. (Schendler, 2009, p. 3)

The bad news comes in the form of a challenge: How fast can we make these beneficial changes happen? Because sustainability, while accelerating, is still lagging behind the growth curves of the problems it is trying to solve. And every single day of delay has a stark cost. Sometimes the losses are incremental, such as the disappearance of a panda, or a Bangladeshi family succumbing to the stress of trying to eke out a living against the odds. But sometimes these costs are huge, and sudden: a climate change-driven storm surge overwhelms a city's flood protection and evacuation plan; a whole species finds that it has no suitable habitat left to migrate to. (AtKisson, 2008, p. 21)

Predictions of irreversible climate change, the urgency of learning low-carbon lifestyles, the unacceptable disparity between global rich and poor, all demand there can be no further procrastination. We might fear that achieving sustainability is impossible. Yet, given that in our own various ways we have all contributed to the problem, perhaps our shared steps in a new direction might lead us on a more constructive path. Learning institutions are uniquely well placed to represent an attractive and positive set of values aligned with humans' best aspirations, and to rise to our most demanding, and potentially most rewarding remit: of helping learners develop the skills to survive and thrive in the challenging conditions of the twenty first century, and contribute to a more sustainable future. (Phillips, 2009, p. 214)

9.3 Is sustainability the answer?

Sustainability is about creating a world where everyone has the ability to meet their needs now and for future generations. At the rate we are consuming, polluting and being unequal, this is not yet a reality. Sustainability means changing the way we think: from competition to collaboration, scarcity to abundance, fear to trust. It means understanding the natural laws and principles for how the natural world works so that we can live within them.

This barrier appears in comments such as, "It's too late" or "Sustainability is a crazy dream", and knowing the scale of the problem it is not hard to understand why one might take a negative view.

A "We're doomed" sentiment represents the pessimist's position. This can be a comforting get-out-of-jail card: I glibly accept our fate, am absolved from any responsibility and can go about my business as usual, like the frog in a pot of water on a hot element not noticing the first bubbles on the bottom.

However, unless the person saying it is genuinely panicking, they are probably not expecting catastrophe in their lifetime or in that of their children. They may be repeating a platitude in order to shut down discussion of something they are not interested in. The human capacity for averting or recovering from disaster should give hope. Evidence of bungled responses (although they are hopefully more the exception than the rule) might give some cause for hopelessness. We are indeed all doomed if we think we are all doomed, because such thinking will sap the motivation needed to developed photocopyable nanotech solar cells, or to devise a clean way of burning coal or to rejuvenate traditional low-impact agricultural techniques.

AtKisson sees recognising the challenge as a guiding principle:

> We have no idea, in most cases, where these hidden triggers and 'tipping points', as they are increasingly referred to, lie—on which specific day is the action deadline to avoid the loss of a species, a catastrophic flooding event, the unnecessary death of a child. If we are serious, if 'caring' is to mean anything, we have no ethical choice but to do our best to make the sustainability transformation happen faster ... and faster and faster. The real basis for hope lies in our willingness to take on this challenge—this responsibility—as one of the central guiding principles in our lives. (2008, p. 21)

Schendler likens the challenge to a boxing match:

> We have scouted this climate problem to death. Yes, we are frightened by the immensity of the undertaking. But this is the opportunity of a lifetime, maybe of a species. The prospect of trying to solve this problem is beyond daunting. It's as if you'd been invited to go into the ring with Muhammad Ali, in his prime, for a fifteen-round bout. The obvious response is: 'No thanks.' But with climate change, you have no choice—someone has a gun to your head and you've got to fight. So what do you do? One option would be to cower in the ring and let

Ali pound you until you die of organ failure. But another approach might be to go for it: bob and weave, dance and waggle, keep your right up, duck and feint—give it your best shot, maybe even have fun … And at least there's a prayer you might get lucky and knock him out. (2009, p. 43)

9.4 Do we even understand what sustainability means?

Perceptions of a lack of definition for sustainability can act as a major barrier to integrating EfS into practice:

> Occasionally, the contestation around such issues as 'definition' rather than core principles can appear like obfuscation or a displacement activity guaranteed to ensure that transformative action is deferred. (Springett & Kearins, 2005, p. 144)

However, as Wals and Jickling (2002, p. 226) note, "it is no use crying over vague definitions". Instead of seeing the ambivalent nature of the concept of sustainability as an impediment, the very vagueness can be an opportunity if it is used as a starting point or operational device to exchange views and ideas. Instead of a tightly defined metric, the very nature of sustainability should be seen as having "many faces and features". There are, then, several responses to this barrier:

1. There are many strong definitions for different purposes and situations.
2. There are core concepts.
3. It is inherently uncertain and contestable, especially in the way it is interpreted by different disciplines.
4. We can use the uncertainty to our advantage.

The concept of sustainability has deep roots, born from the 20th century realisation that human activities were endangering future life on the Earth. Definitions of sustainability vary widely, from a strong environmental focus:

> Sustainable development—improving the quality of human life while living within the carrying capacity of supporting ecosystems (International Union for Conservation of Nature, United Nations Environment Programme & World Wildlife Fund, 1991, p. 10)

through economic approaches:

> Sustainable development: The amount of consumption that can be sustained indefinitely without degrading capital stocks, including natural capital stocks (Costanza & Wainger, 1991, p. 8)

to a broad global view:

> Sustainability—The ability to meet present needs without compromising those of future generations. (World Commission on Environment and Development, 1987, p. 43)

Current definitions generally include three components—environment, society and economy—along with the recognition that the three areas are intertwined, not separate. There may be a lot of disagreement, but if people are genuine in their concern, then basically they are all heading in the same direction: the maintenance or enhancement of natural and social systems to ensure the enduring coexistence of the natural and cultural worlds.

Sterling (2004a) has argued that sustainability, rather than a measurable outcome, is an emergent quality arising from sets of relationships in a system, whether viewed at the macro or micro scale. In this way it is rather like justice or health, as Holmberg and Samuelsson (2006) have argued. Despite all the attempts to make the concept more operational, it is clear that it cannot be exactly defined— and it should not be. It is an ever-evolving concept:

> It can be compared with the concept of health—as health cannot be defined in precise terms either, and yet, everyone has an idea about what health is and health is important for everyone. When we meet each other we often ask: How are you? Sustainable development can be seen as the health of societies and the planet. If we are concerned about the present development and whether it is sustainable we instead ask each other: how are we? (Holmberg & Samuelsson, 2006, p. 8)

It is clear that sustainability cannot be defined without reflecting on values and principles. As a result, any discourse about sustainability is essentially an ethical discourse (Bosselmann, 2008). So is sustainability just ethics rebranded? What is new about it? Is sustainability different?

Amidst the complexities of ethical literature, I like this simple observation from Koehler and Som (2005, p. 16) in a discussion of pervasive computing: "sustainability is the extension of the ethical principle of justice in space and time".

Sustainability, a mindful (*response*-able) approach to our interaction with our environment and its interdependent living and nonliving constituents, is driven by a sense of responsibility for ourselves (those alive now and those yet to be created) and other organisms, and a respect for the systems in which we live. To consider the welfare of others and enable them to thrive also allows us to have a more meaningful existence. Unsustainability has been the product of imprudent (unethical) accumulation and waste on the behalf of a minority, who have no more right to the squandered resources than the less affluent majority. An essence of sustainability is a common ethical foundation (AtKisson, 2008), and this is made explicit in the Earth Charter (Newman, 2009).

9.5 It's their problem, not mine

This barrier, abdicating responsibility for environmental problems, tends to arise in the face of seemingly insurmountable and distant problems, such as China's coal-fed power stations. It arises, at least in part, from a sense of disempowerment, a belief that it's not worth me doing anything because of other, much bigger problems that I can't influence. Sometimes, too, there is a sense of competition—that we're not going to sacrifice any potential advantage while our competitors are not playing ball.

Hamilton and Denniss (2005) make it clear that we need to take responsibility for our own consumption—the coal plants in China are providing our stuff—and much of it is unnecessary.

There is no doubt that the question of sustainable development in the so-called developing world is vexed. Should we be trying to stop consumption growth in the third world? Is it reasonable to set a much lower target for people to aspire to than that which we have long since surpassed? We in the developed West need to be careful not to make assumptions about the aspirations of other people.

9.6 But what about the bottom line?

I am doubtful about claims that connect green with short-term business benefits. Healey (2009) found that just 12 percent of the 419 business technology professionals surveyed would be willing to pay more for a greener product. Healey attempts to go beyond the decisions that have clear monetary return on investment, but, perhaps not surprisingly, there's not much there:

> ... hard return on investment was rated as the most important criterion for evaluating green alternatives ... and, when it comes to purchasing requirements, the top ranked green feature was power consumption; the lowest ranking was whether or not a sustainable manufacturing process was used.

This is not because of a lack of personal understanding of environmental issues, nor a lack of personal commitment. Over 90 percent agreed that they seek environmentally friendly alternatives in their personal lives, yet few are doing much at work.

Why is green IT such a nonstarter? Partly because it's difficult, partly because of inconsistent policy and direction, partly it stems from a lack of standards and partly it is due to self-serving green IT messages from current vendors. Mostly, though, it is because being sustainable is not highly rated in the business environment, except perhaps as a marketing tool.

IS THERE REALLY A PROBLEM, AND IS SUSTAINABILITY THE ANSWER?

Others have reported similar findings to Healey. IBM and the New Zealand Business Council for Sustainable Development (2008) surveyed New Zealand businesses and reported that 94 percent of people recycle at home for environmental reasons. At work, however, only 11 percent of the same people recycle. The same applies in environmental transport decisions: at home, 40 percent; at work, only 6 percent. At home, it seems we are starting to care: an encouraging 72 percent of us report environmental considerations affecting purchasing decisions. Yet at work only 28 percent report any environmental consideration in purchasing at their workplace. We seem to have reached a position where caring for the Earth is something that is acceptable, even expected, before school and in the evenings and weekends, but is not something we do from nine to five. Why is this?

Presumably people know that any hope we have to make the world a better place in which to live and work depends on making changes at home *and* at work. Changes at work will have so much more impact than those we can make at home. While I applaud people who recycle their compost at home, it starts to pale into insignificance if they spend their days driving thousands of kilometres primarily focused on selling more cheese than a competitor (who is doing exactly the same). There has to be a better way.

Does this suggest a significant change in the way we do business? Possibly. It is the basis of the sustainable practitioner's approach. It suggests that while every bit counts, sustainability in the workplace needs to be more than recycling in the lunchroom. To progress to eco-enterprises will take a much wider incorporation of sustainability into business and a much wider understanding of sustainability. This means a focus on holistic thinking rather than checklists. You are not going to be measuring ecological friendliness just by watching the power meter.

Yet in the attempt to make sustainability real and make it accessible to business, we have lost the connections. Reducing everything to simple dollar equivalents blows away the inherent complexity. Hallstedt (2008) points out that sustainable decision making requires a consideration of a decision's impact in degrading the social system. This includes abuse of authority (enforced labour undermining labour unions, etc.) and abuse of economic power (nonliveable wages, exploiting investments, etc.). If it was shown that your preferred supplier was using enforced child labour, would you be prepared to change suppliers, and how much more would you be prepared to pay?

9.7 This stuff is just leftist nonsense

Schendler (2009, p. 115) succinctly describes the effect that the origins of the environmental movement have had on the perception of sustainability today:

> In a way, modern environmentalism, which is pragmatic, businesslike, collaborative, and climate-focused, has been hamstrung by historical environmentalism, which was often shrill, exclusionary, irrational, and micro focused. Being mischaracterised as a tree-hugger is something that makes my job, and the jobs of others in my field, much more challenging than it would be otherwise.

Sustainability cannot help but be political when it forces an examination of the balance between individual and universal rights:

> If protecting individual health relies on exploiting resources to the extent that others are denied the opportunity, then it is unsustainable. So global sustainability is dependent on assuring the rights of many, possibly at the expense of some individuals, and the challenge of balancing these constitutes another significant barrier to achieving sustainability. (Sibbel, 2009, p. 73)

However, encouragingly, a *New York Times* poll found that 63 percent of those polled agreed that protecting the environment is so important that continuing environmental improvements must be made regardless of cost, and that the importance of this has reached across party lines, with a majority of both Democrats and Republicans in support (Dumaine, 2008).

CHAPTER TEN

Is it the responsibility of higher education?

10.1 Is it really up to us to do this?

Resolution and amelioration of complex global problems in the longer term ... falls to the professions and, through the education of professionals, to educational institutions. (Tomkinson, 2009, p. 165)

The dominating motif of Newman's (1873) 'idea of a university', with its moral ideal of the disinterested pursuit of truth that combines human understanding with knowledge, has long since faded away, like the dominance of the classics. (Cullingford, 2004, p. 13)

Governments and groups of higher education institutions have made many international declarations on EfS,[4] but, surprisingly, the notion of developing more sustainable physical operations on the university campus does not seem to have been a priority for the majority of them. Instead, emerging themes in the declarations are of lifelong learning, moral obligations, public outreach and ecological literacy.

---

4 Wright (2004) and Nicolaides (2006) chart the history of development: Stockholm, Tbilisi, Tallories, to Lüneburg, Rio, Halifax, Copernicus, Bonn, etc.

BARRIERS TO INTEGRATING SUSTAINABILITY INTO ACADEMIC PROGRAMMES

In 2004 the United Nations declared 2005–2014 the Decade of Education for Sustainable Development, recognising the pressing need for education for sustainable development as "a life-wide and lifelong endeavour which challenges individuals, institutions and societies to view tomorrow as a day that belongs to all of us, or it will not belong to anyone" (United Nations Scientific and Cultural Organization, 2006). The United Nations (2004) argues that universities and higher education institutes can contribute to sustainable development in several ways:

- by giving sustainable development a place in all university curricula and educational and research programmes
- by playing an important role as local knowledge centres for sustainable development
- by making sustainable development a leading principle in their own logistics and managerial processes. (United Nations, 2004, p. 2)

The call for sustainability in education is not new. Nicolaides (2006) reviews a long history of international efforts to promote links between sustainability and education. In 1977 UNESCO's Tbilisi conference set out objectives for environmentally conscious educational institutions. The 1987 Brundtland Commission was followed by UNESCO calls for strategies on environmental education. The International Association of Universities signed the Tallories Declaration in 1990, with its aims of all university graduates being environmentally literate and the development of curricula for an environment that is sustainable.

In 1992 the Rio Earth Summit *Agenda 21* identified four major thrusts of education for sustainable development:

- basic education must focus on imparting knowledge, skills, values and perspectives that encourage and support citizens to lead sustainable lives

- reorienting existing education at all levels to address sustainable development
- developing public understanding and awareness of sustainability
- training and the development of specialised training programmes to ensure that all sectors of the workforce have the knowledge and skills necessary to perform their work in a sustainable manner. (United Nations, 1992, Section IV, Chapter 36.4)

In 1993 the European universities adopted the *University Charter for Sustainable Development* (known as the Copernicus charter) and *Agenda 21* was reaffirmed at the Johannesburg Earth Summit (the Ubuntu declaration). More recently, all programmes have come together under the UN Decade of Education for Sustainable Development 2005–2014, and operationalised in the Global Higher Education for Sustainability Partnership (GHESP).

Cortese reminds us of the importance of EfS for all professionals:

> Unless higher education responds quickly to ensure that all of their graduates, regardless of their fields of study, are environmentally literate, then it is unlikely that our future leaders will demonstrate the analytical thinking, the will or the compassion to adequately address complex issues such as population, climate change and social equity … Environmental specialists alone will not help us move toward a sustainable path. A compartmentalized approach further reinforces the assumption that environmental protection should be left to environmental professionals. All humans consume resources, occupy ecosystems and produce waste. We need all professionals to carry out their lives and activities in a manner that is environmentally sound and sustainable. (Cortese, 1999, p. 6)

In New Zealand, the *Tertiary Education Strategy 2007–2012* introduced sustainability as a key direction from the first line:

> Tertiary education and research underpin the realisation of New Zealanders' goals and aspirations and the sustainable development of New Zealand's economy and society. (Tertiary Education Commission, 2007, p. 4)

Tertiary education has been accused of contributing more to the problems than to the solutions (Lautensach, 2004; Orr, 1994):

> The destruction of the planet is not the work of ignorant people. Rather it is largely the result of work by people with BAs, BScs, LLBs, and PhDs. (Orr, 1994, p. 7)

The argument is that because the large majority of people in crucial decision-making positions are likely to have tertiary qualifications, the greatest amount of harm is done by people with higher degrees. This is not a recent phenomenon. Once could argue that unsustainability directly tracks the development of education over 200 years. So why should more education make a difference?

> The educational system, of course, is at the heart of our current unsustainable society, being both its product and its creator. It is probable that the people taking the decisions that determine the trajectory of our global and local development are generally themselves the most successful products of what we might agree is a flawed educational system. (Phillips, 2009, p. 209)

This flawed history, though, is not an irresolvable problem. Current education systems might be the problem, but future education can be the solution. Education systems reflect the concerns, priorities and values of the society in which they operate. Schools, universities and polytechnics endorse waste and consumerism by their selective silence on these issues. These institutions may produce the graduates who go on to engage in enterprises such as marketing cigarettes or real estate ventures that pay scant attention to environmental impact, but they also produce the nurses, the teachers and the traditional furniture makers. However, education is not always silent on the shortcomings of our economic, technological and social development. Many disciplines in higher education (social work, law, philosophy, gender studies) deal with issues of justice and equality, and others (geography, science, ecology, political science, peace studies) with human impacts on the natural or cultural world. Education acts as society's conscience. In higher education people can, if they wish, develop the skills of critical thinking, of communicating ideas and of pursuing the truth without the coercion of corporations or governments or others who have an agenda.

10.2 Shouldn't higher education be values free?

Despite widespread acknowledgement that a values-free classroom is impossible, this barrier is commonplace, but misguided:

> No education is value-free. For example, the topics that teachers select and the content of their school's curriculum are all based on value judgements about what matters most in education. (Parliamentary Commissioner for the Environment, 2004, p. 43)

EfS needs to be open about its values-based approach because it contrasts with the views of many individuals and sectors of society who believe that education can and should be value free. Some people fear that values are somehow linked to indoctrination, despite the fact that education systems are already shaping people to think and act in certain ways. Ultimately, EfS requires people to think critically about and reflect on their own values and the values embedded in the institutions that surround them in order to decide what values a society, and the different groups within it, may wish to adopt.

The strongest proponent of the value-free stance is Stanley Fish (2008) in *Save the World on Your Own Time*. Fish starts by asking two questions: "What is the job of higher education?" and "What is it that those employed in higher education are trained and paid to do?" The answer to both is: to introduce students to bodies of knowledge and methods of enquiry, and to equip those students with analytical skills to move confidently in that field. Fish (2008, p. 19) says, "Nothing else, nothing more."

Fish is highly critical of both academics and institutions that see other roles for themselves in producing socially and ethically conscious graduates. Instead, Fish argues, education should be values free, with issues "academicised—detached from the context of real world urgency" (p. 27). This, he argues. will change the inclination of students to change the world, into an urge to understand. For Fish the crucial question is not "What do you think?" but "What is the truth?"

Newman takes quite a different position. Acknowledging that it is not the role of higher education to teach values explicitly, as to do so could attract criticism for imposing ideologies on learners, he says:

> But if all mention of values is expunged from education then this leaves little choice but for learners to draw their values from the unsustainable society around them, or from the values latent in the 'hidden curriculum' of their educational institution.
>
> Values such as intellectualism, competitiveness, rationalism, technical instrumentalism, reductionism, and scientism may well be hidden within the presuppositions of the curricula of learners' institutions, their textbooks, formal lectures and assessment strategies.
>
> Values reflection is one way out of this dilemma. Rather than having values imposed on them, learners reflect on the dominant values of society and their institution in the context of the changes that are occurring in the world around them, and ask themselves whether these values are now outdated, or even dangerous. (2009, p. 99)

AtKisson (2008) is another for whom values-free education is impossible. As he succinctly puts it, "First, you have to care."

10.3 What does it mean for links with community and business?

Let me tell you a story. My colleague Brian has a jacket for a favourite race-car team which has fascinated me since he got it. I wondered what sort of negotiation went on that resulted in a smoking company and a quit-smoking company jostling for position on the same arm? Brian's jacket was on my mind when I saw a letter from the leader of one of Dunedin's organic garden groups following a talk I gave to Sustainable Dunedin City, where I mentioned that we were talking with several Dunedin businesses (one in particular) about the possibility of corporate sponsorship. This upset my correspondent, who wrote:

> I don't know what kind of sponsorship they are talking about, i.e. if it is cash sponsorship or materials sponsorship, but either way I am pretty sure that accepting any kind of sponsorship from such a business is inherently unsustainable. I understand that sometimes you have to take what you can get. And it depends on how fussy/holistic you want to be with the overall model of sustainability.
>
> I don't know that much about economics, but I have a rough enough picture of it. The reason that taking sponsorship from [Dunedin business] is unsustainable, is that the particular form of capitalism that it is functioning within, has a slow but inevitable outcome: the concentration of wealth in the hands of a small elite of rich people, oligopolies in the market place, crushing of small independent businesses, etc etc.
>
> This may sound radical, but it would seem that this system is destined to collapse which means that it is not sustainable. Constant growth is not sustainable, which is what capitalism requires. And even if it was sustainable, it's just not nice! Hah.
>
> And what will they get from the sponsorship? Advertising. They aren't even seeing it as sponsorship. They're seeing it as marketing. If they can have their logo associated with an amazing, organic, wholesome community initiative toward sustainability, then people will assume that [Dunedin business] supports and perhaps even abides by those ideals itself. But alas, it does not. [Dunedin business] and its buddies are symbols of one of the several significant problems facing western civilisation at this point in time.
>
> What is the funding for? Why is so much money needed? Gardening really isn't that costly. I'm advocating a more DIY approach requiring minimal monetary input. Surely the polytech is contributing some funding itself, from student fees? Is this not enough?
>
> Anyway I'm not trying to sound like I'm getting on a high horse or anything because I don't know much about the project and I realise

that funding can be hard to come by. But I just thought I would put up that angle of perspective for you to consider.

Here's my reply:

Thank you for this ... I understand your point of view here and agree with much of it. There are clearly potentially strong feelings in the community, but yes, we do need to engage with Otago's businesses.

You are right that Otago Polytechnic should be investing in Living Campus, and indeed it is. We need to be careful though, while we see Living Campus as being very beneficial for learning, it is not as obvious as, say, employing a lecturer—in the short term at least investment in these initiatives could be seen to be taking money away from direct teaching and learning. In this regard, it is important to remember that the Living Campus is 'more than a garden'—it is also a centre of engagement in learning around sustainability. This is going to take effort and investment in integration into learning, development of teaching resources and a heavy focus on interpretative and interactive resources.

The Living Campus is also a busy campus—something like 14,000 people come through the door each year (and that's not counting the passing foot traffic, hockey players or skateboarders). We do not have the luxury of a home garden—everything we do is scrutinised by health and safety, disability access, building consent requirements, etc. We are carefully treading a tightrope between an organic feeling and a high use outdoor museum. This is, unfortunately, not cheap.

So, yes, we do need the money. Perhaps more important than this, though, is we need the engagement of the whole community—including the businesses. While we may not like some aspects of what they represent, the fact remains that they are here. Most importantly, the hardware/gardening stores are where most people are getting their materials, resources and knowledge. To get the people, we need these stores to move to a sustainability message. To misquote someone, we need them in the garden, not outside it.

Your email has prompted me to look for advice. The best I can find is the Ethical Consumer website. This UK group recognises, as you do, that 'Sponsorship is not philanthropy. It is a mutually beneficial business partnership. They suggest a code of ethics that is used to 'share a common understanding of the charity's ethical values'.

I think we need to write a set of guidelines for association with Living Campus. You are right, people will want to 'associate with the amazing, organic, wholesome community initiative toward sustainability'. And you are also right that other people will assume some sort of alignment of values.

I've spoken about this with our business relationship manager. He suggests a softly softly approach, the retail sector is not strong at the moment, too big a stick and we'll lose both the opportunity to engage and the possibility of money. We have quite a big stick in our procurement policy which says that at Otago Polytechnic purchasing decisions, at whatever level these are made, are expected to take into account both financial and sustainability issues, and to contribute towards meeting the polytechnic's objectives in the area of sustainability ... and that a sustainable approach to purchasing means taking into account social, environmental and broad economic factors when meeting the polytechnic's needs for goods and/or services. To achieve this, the polytechnic will support suppliers which are socially responsible and have adopted ethical practices, which avoid unnecessary consumption, which select goods and/or services that have a lower environmental impact across their life cycle than competing goods and/or services and whose work practices demonstrate innovation in sustainability.

We clearly need to build relationships with the businesses in our community as well as groups such as your own. By working with the businesses we can contribute to their progress down this sustainability pathway. Would you be able to help us with this, perhaps a short working party to write a Living Campus Code of Ethics?

10.4 Is this what students sign up for?

In my experience it would be a rare student body that didn't respond positively to an institutional sustainability initiative. In most institutions I have visited the student union and student groups in general are a long way ahead of the institution in terms of progressing sustainability.

The primary and secondary sectors do a remarkable job of preparing students. The Programme for International Student Assessment (PISA) and the Organisation for Economic Co-operation and Development's (2009) *Green at Fifteen? How 15-year-olds perform in environmental science and geoscience in PISA 2006* report suggest a cohort of environmentally competent young people. They arrive at tertiary education aware of sustainability and environmental issues, but may well not have reflected on them in the context of their chosen discipline, nor have they learnt the appropriate behaviours or tools that discipline brings to bear on these issues. That is our job.

10.5 Surely it's the role of the media rather than of higher education

The media have an important role to play in sustainability:

> Given that a great deal of our understanding of the environment—its habitats and peoples—is derived from new and traditional media, it is imperative that any notion of sustainability literacy is connected with ongoing work on media literacy. (Blewitt, 2009, p. 111)

Osbaldiston and Sheldon (2002, p. 46) observed that the sustainability message in the media is overwhelmed by messages going in the other direction:

> For every 60- or 90-second spot on television about global warming, there are probably thousands of other commercial messages that glamorize fuel-inefficient vehicles, air travel ...

The representation of sustainability in the media provides useful material for teaching, especially in critical thinking. Sustainability intensity metrics are considered the next big thing in sustainability management and reporting. They attempt to give an indication of the "good bang for the bad buck". Used properly, intensity metrics make more transparent the balancing decisions that face sustainable practitioners. Unfortunately, they can also be used to obfuscate, or "greenwash".

Makower (2009), for example, reports "good news ... each year we are using less paper", but this is on the basis of "paper intensity" or the thousands of tons of paper used per billion dollars of GDP. In reality the amount of paper used per person in the US has stayed about the same (at around 302–340 kilos per year).

So where did Makower's *State of Green Business* report go wrong? The big clue is the cited source of the data, the American Forest and Paper Association (AF&PA). I have trawled all over the AF&PA website (http://www.afandpa.org/) looking for these data. Unfortunately, and dealing a blow to their transparency and credibility, their "statistics" links to a bookshop with no sign of data (even if I was prepared to pay for it). At first glance, the reports on the website look like an advertisement for the Greens, with images of trees and waterfalls surrounding headings of "clean" air and "clean" water. But when you actually read the text, the extent of the greenwash becomes apparent: the association proudly lists the environmental measures they have recently actively opposed. Despite the tag line of "Improving tomorrow's environment today", this place doesn't seem a credible source for a state of green business report.

10.6 They don't need this, it's all over the Web

> The Web has changed everything. And the social web is empowering a new class of authoritative voices that we cannot ignore. (Solis & Breakenridge, 2009, p. 1)

The impact of Web 2.0 might provide new ways of engaging with students, but we do still need to integrate sustainability. Ernesto van Peborgh and the Odiseo Team (2008) successfully argue that Web 2.0 has had a major impact on how we think and how we communicate, and that to be successful, businesses should adopt an Enterprise 2.0 mindset. The authors characterise Web 2.0 as a "series of disruptions" and argue that the behaviour change—to an interactive and participative medium—is more important than the technology itself. Although van Peborgh et al. don't manage to integrate Web 2.0 and sustainability, they do discuss how "emergent collaboration" is changing business:

> The medium, or process, of our time ... is reshaping and restructuring patterns of social interdependence and every aspect of our personal life ... Everything is changing: you, your family, your education, your neighbourhood, your job, your government, your relation to 'the others'. And they're changing dramatically. (2008, p. 142)

Van Peborgh et al. conclude that there is a global crisis, which needs a new paradigm, although we are not yet sure what this is. Enterprise 2.0 has some useful characteristics—respect, openness, creativity, new technologies—for whatever this new paradigm might be, and some of these align well with the characteristics desirable in sustainability. The Web and other new media are important in addressing sustainability. The trick for educators will be to use them to engage students in being critical and creative thinkers, as Blewitt argues:

> The key to success is harnessing new media to capture imagination and interest and transform understanding, values and knowledge. The critical understanding of sustainability communication and new media can best be achieved through an informed engagement with the emerging technological affordances evident in many areas.

> We are all media practitioners now. We need to be sustainability practitioners, too. (2009, p. 112)

CHAPTER 11

I can't see this working in our institution

11.1 What happened to academic freedom?

Academic culture often perceives external threats to academic freedom. This is a long-standing tradition, which has its origin in the resistance of academics to control by political authorities. However, viewed as an opportunity, academic freedom gives universities and other tertiary institutes the power to choose what they teach and research—no-one is stopping anyone from integrating EfS. Scott and Gough (2006, p. 91) describe one vice-chancellor's response to the UK Higher Education Funding Council (HEFC) publishing a sustainable development strategy:

> It is not the job of universities to promote a particular political orthodoxy; it is their role to educate students to examine critically policies, ideas, concepts and systems, then to make up their own minds. The Funding Council should support that objective, including, from time to time, telling the government that the university curriculum is none of its business ... Trust me; it (the strategy) is not harmless. It is one of the most pernicious and dangerous circulars ever to be

issued. It represents the final assault on the last remaining freedom of universities.

The HEFC backed down on the plan to link funding to sustainability, and the barrier remains.

> However it must be clear that the call for sustainable development is not prescribing any specific world order. It only calls scientists to address the global problems that threaten the continuity of civilized society, and does not prescribe specific solutions. (Mulder & Jansen, 2006, p. 71)

> In the long term, sustainable university development can only take place successfully if it is presented to universities as an *optional possibility*, as universities have effective defensive and relativising strategies for dealing with unwished for top-down regulation. (Adomssent, Godemann, & Michelsen, 2007, p. 389)

11.2 What about the sustainability of our student intake?

This barrier is often expressed in statements like "I'll get on to that when our department is economically sustainable" and "The department is under review, we'll do it after that."

This is an unfortunate use of the term *sustainable* because it shows a lack of understanding that the viability of some courses and programmes might be improved through a revitalisation involving the integration of sustainability. Sustainability can be an important driver in that transformation.

For example, despite strong career prospects, computing education continues to suffer low student intakes, especially of women (Computing Research Assocation, 1974–2010). "Something is wrong with perceptions of computing", can be heard around the halls of every campus and computing conference, whereas actually the truth may be that something is wrong with computing itself. Benyo, White, Ross, Wiehe and Sigur (2009) go some way to uncovering what lurks behind this image problem and suggest that what is wrong is the very nature of computing.

They confirm that there is a significant gender gap among US college-bound students in their opinions of computing as a possible college major or career. Boys see computing as a very good choice for study (74 percent) and career (67 percent), views not shared by girls (32 percent and 26 percent). Benyo et al.'s survey went further, however, and asked the students what was important to them in their careers. Here the differences between the sexes become even more apparent:

- 64 percent of boys rated "being passionate about your job" as extremely important; 78 percent of the girls felt the same
- earning a high salary rated as extremely important to 50 percent of boys, but only to 39 percent of girls
- "having the power to do good and doing work that makes a difference" rated extremely important to 56 percent of the girls, in comparison to 47 percent of the boys.

Furthermore, "having the power to create and discover new things" and "working in a cutting-edge field", two of the things used to sell computing as a career, were way down the list of career preferences. This shows that computing educators are appealing to a tiny minority by using this strategy.

The respondents were also asked to rate a selection of potential computing marketing statements. The traditional messages, "computing opens doors" and "computing puts you in the driver's seat", were favoured by boys, but a quite different message, "computing empowers you to do good" topped the girls' rankings (see Table 3).

Table 3 Survey results from US teenagers shows a considerable gender gap in responses to potential career branding phrases. The table shows the percentage of each gender who favoured different proposed computing marketing statements.

	Total	Boys	Girls
Computing puts you in the driver's seat. Why merely create a MySpace page when you can create the next MySpace? Computing gives you the power to imagine new languages, new worlds and new ways of improving our lives by putting better ideas into actual practice in our communities.	42	48	35
Computing opens doors. With eight billion computers in the world, just about everything depends on computing today. From transportation and energy to video games and space exploration, few careers enjoy so many real-world applications and few open as many doors as computing.	39	50	29
Computing empowers you to do good. With computing, you will be able to connect technology to your community and make a world of difference—reducing energy consumption, improving health care, enhancing security, reducing pollution and advancing learning and education.	37	36	38
Computing brings people together. Computing connects you to a world of smart, creative people who share a passion for new ideas, new inventions and new solutions that impact our lives, our nation and our world.	31	36	27
Computing is achievable. You don't need to love math and science to be a computing leader. But you do need to have a passion for solving problems and for working with others to figure out how to make all this computer stuff work better.	26	32	21
Computing puts you on the cutting edge. When you choose computing, you are plugged into the latest innovations, and you understand like few others how connected we are as businesses and people in today's digitally driven economy.	26	36	17
Computing calls for creative problem solvers. Computing requires you to know what you want to solve and how you want to solve it. That takes creative thinking, a powerful imagination and real collaboration.	24	30	18

Source: Benyo et al. (2009)

11.3 Sustainability is not recognised by our funding model

That sustainability is cross-disciplinary, and therefore not recognised by the funding model, is a challenge for higher education. There are, however, many examples of successful teaching and research centres that bridge disciplines, including child health, energy and health informatics.

A separate, stand-alone course has the advantage of being able to focus explicitly on sustainability. The danger of a separate course is that it is just that: a separate course. The worst outcome is that students see sustainability as a distraction from learning what they are there for. Integration gives the opportunity to provide context. It has attendant dangers too, of course, primarily that it can become "everywhere and nowhere".

In describing Education for Sustainable Development and Global Citizenship, the Welsh Assembly Government (2007, 2008) explicitly emphasises that it is not:

- a separate subject
- a series of discrete concepts or topics
- confined to the classroom
- the responsibility of just one teacher in the school
- about transmitting a set of answers to learners.

11.4 Sustainability is not valued by my institutional hierarchy

In my role as chair of the national group promoting sustainability in tertiary education (Sustainability in Tertiary Education in New Zealand—STENZ), I have written to the head of every university, polytechnic, institute of technology and wānanga in New Zealand. About 80 percent replied and every single one of these was supportive. Most pointed to their institutional mission statements, which include words to the effect of sustainability (or at least citizenship, or bettering the community). A few mentioned academic freedom, noting that they cannot direct their staff to include a particular subject area.

Despite this (admittedly anecdotal) support from institutional management, the institutional hierarchy barrier is regularly mentioned. It is different from management not being supportive: it concerns a perception that senior management have actually prevented the integration of sustainability in teaching. In all but two cases where people have mentioned this barrier, the issue has boiled down to a misunderstanding over directives. In only two cases that I know of there has been explicit instruction *not* to teach sustainability. From different institutions, both have involved a new senior manager telling a lecturer to "get that climate change stuff out of your course on computing ethics". In one case the lecturer resolved the issue by inviting his dean to contribute to a class panel discussion on challenges facing computing over the next five to 10 years. In the other case there needed to be continuing discussions about academic freedom.

11.5 How does this fit with all we're required to do?

This barrier is faced by disciplines teaching to strict professional requirements, often with an external exam (such as nursing), or to a prescribed curriculum (National Certificates). People in such disciplines voice concern that there is no time to fit more in. However, the key is to realise that professional requirements specify a minimum, and that inclusion of EfS need not be onerous.

An approach that can help resolve the tension is to influence the national bodies that set the national qualifications. We have had much success with the NACCQ, which sets the curriculum for several certificates in New Zealand.

Otago Polytechnic teaches the National Certificate in Horticulture (Advanced) (Level 4). Although the National Certificate is a prescribed programme, the programme itself is determined by the polytech and described in a programme document. The school management included the following in the graduate profile:

> Graduates will be able to apply their technical knowledge of sustainability to critically evaluate the environmental impacts of their actions and continually improve their practices to match current industry sustainable practice. (Otago Polytechnic, 2009a, p. 4)

The graduate profile is an Otago Polytechnic statement rather than a National Certificate one. The restriction is that the school can still only assess the unit standards—as soon as it adds other units it ceases to have the status of a national qualification—but the school is happy to integrate sustainable practice into the programme.

The NACCQ sets the curriculum for several certificates in New Zealand. In July 2008 it adopted the following policy:

> Computing and IT underpins every sector of society as a pervasive and influential discipline with global impact. The vision is that our graduates, our practitioners and our academics understand the concepts of social, environmental and economic sustainability in order for them to evaluate, question and discuss their role in the world and to enable them to make changes where and when appropriate.
>
> Moreover, computing educators must take a lead in sustainability so that computing practitioners can be encouraged and supported to promote sustainable use of technology. This can primarily be achieved by the fostering of sustainability as a core value of computing education.
>
> Creating a philosophy of Computer Education for Sustainability will be enhanced if undertaken within a context of institutional operational practice. We will then be seen to be modelling good practice. (National Advisory Committee on Computing Qualifications, 2007, p. 5)

Subsequent reviews of curriculum documents have seen EfS integrated into many courses. These courses are taught in more than 20 tertiary institutions.

11.6 We'd rather the students were critical thinkers

This is not a dichotomy. Sustainable practitioners are critical thinkers who are also caring, creative, systems thinkers. Sterling (2004a, 2004b) says that the emphasis of EfS must be on the quality of learning and on building individuals' capacity to think critically, systemically and reflectively, rather than on encouraging particular social or environmental outcomes. Knowledge about environmental issues and sustainability is constantly evolving, and it will only be by developing students' critical skills in evaluating information that we can really equip them to be sustainable practitioners. Holmberg and Samuelsson make this point:

> Sustainable development brings many challenges to the universities. Universities, with their core values of scepticism, curiosity and freedom of speech, have a profound role to play in developing students' qualities to cope with uncertainty, poorly defined situations, diverging norms, values, interests and reality constructions. (2006, p. 8)

11.7 We're doing it already in our own way

Well done. So why is this a barrier? If it is only a small part of what you do, which perhaps you see as a box-ticking exercise, then perhaps a rethink is in order. If you are genuinely pleased with what you are doing then please share it widely with others so that they can learn from your experience.

There is a danger in thinking that because your programme is teaching one aspect of sustainability, this transfers to a more holistic understanding. There is a greater danger if this aspect is only tangentially related to sustainability.

It would be a mistake to attempt to teach every aspect of sustainability. Not only is there not room in programmes to do so, but the goalposts keep moving. Next decade's issue might not yet have arisen. This is not new, of course. Many disciplines (computing especially) face this problem. The appropriate response is to teach to a framework so that detailed skills learnt in one area can be transferred to another.

It is not entirely clear whether learning is helped by making it explicit that you are talking about sustainability. If you are teaching about energy management, and using light bulbs as an example, does it help to mention that principles of energy conservation underpin much of sustainability? Would that help graduates making decisions about whether to disconnect the power in a house with an elderly person on a breathing apparatus?

> Without the right tools, learners faced with these wicked problems may fall back on the same old inappropriate toolbox with, at best, disappointing outcomes. Given the messy nature of the dilemmas and contradictions facing us there can be no single recipe and no definitive set of tools. (Morris & Martin, 2009, p. 156)

Sometimes it is argued that sustainability properly belongs in a professional practice course. Professional behaviour is indeed a good place for sustainable practice (Love & Love, 2008). Wary of a tendency to dump noncore material in the lap of professional behaviour, I would argue that the professional behaviour course is integral to this process of transforming education, but that this cannot be the totality of the integration of sustainability into a programme.

As a society we have to learn to live in a complex world of interdependent systems with high uncertainties and multiple legitimate interests. These complex and evolving systems require us to think simultaneously about the drivers and impacts of our actions at different scales, and about the barriers of space, time, culture, species and disciplinary boundaries. This means our graduates need skills in systems thinking, an understanding of the connected nature of our socio-ecological system, critical and creative thinking, an ability to act as change agents, an understanding of ethics and a sense of participation and action. None of this is possible without the framework of communication and professional behaviour.

Sustainability should not be seen as an extra subject and should not be confused with being green, or with education about the environment. Instead, sustainability provides a context for learning within and about the student's discipline. This sustainability lens overlaps strongly with professional practice—hence the sustainable practitioner. But, just as we expect professional practice throughout the programme (not just in a professional practice course), sustainable practice must also be threaded throughout the programme.

CHAPTER TWELVE

I want to integrate EfS, but how do I do it?

At this level of agreement the reader believes in sustainability as a response to system degradation, and they agree with the notion of the sustainable practitioner and that it is the responsibility of higher education to prepare people for their careers as sustainable practitioners. They are just not sure how it relates to them or their teaching. To some extent this is a matter of personal responsibility: for some it is an issue of motivation, for others the barriers might be about grasping the relevance, and still others are just unsure how to proceed.

12.1 I can't really see how it fits my discipline

If your subject area has no real-world relevance at all, then perhaps sustainability has no relevance. If none of your graduates apply anything they have learnt to any form of decision making that affects real people, or real systems, then perhaps sustainability has no relevance. If, on the other hand, your graduates do work in areas that involve actions affecting people, or physical or natural systems, then sustainability is a concern for your discipline:

> Any closed definition of sustainability education (assuming it can be achieved to common satisfaction) involves drawing conceptual boundaries. This carries the danger that all other educational policies, theories and practices appear to be outside or beyond these boundaries—and, therefore, actors outside of the boundaries assume or perceive that sustainability education is not their concern. There is a parallel here with sustainable development: either it is an add-on area of theory and practice (that is, a sectoral interest), or—to a greater or lesser degree—it involves all aspects of social, economic and political organization. (Sterling, 2004a, p. 56)

Within disciplines it is entirely possible that there are subdisciplines for which it is difficult to see an application of sustainability. However, it has been my experience that, given this acceptance, many people in subdisciplines which we thought would fit into this category have come forward to share the exciting things they are doing:

> Behind every gold wedding ring lies a genuine gold mine, and the possibility of a massive cyanide spill. Behind a tuna steak is a decimated tuna population. Behind a comfortable car is a strip mine, a hundred toxic chemicals leaching into nature, and war in the Middle East. (AtKisson, 2008, p. 46)

Remember, and this is worth repeating: sustainability should not be seen as an extra and should not be confused with being green, or with education about the environment. Instead, sustainability provides a context for learning within and about the discipline. For me, in my discipline, it is the lens for software engineering. It not only provides relevant practical projects, but also many of the concepts align.

It often seems sustainability is a good thing when it's in someone else's course. We are then in danger of letting it fall between the cracks. There are cases, though, where including sustainability does seem a bit of a stretch. We hear all the time things like, "I teach algorithms—it's completely abstract. Surely I don't have to put it in there?" And, while we can see opportunities for case studies and context, in some cases we agree with these people.

So, how much and where? At Otago Polytechnic we have been exploring a simple matrix to help departments target their integration of EfS. To develop the matrix we have talked with people from a range of departments and asked them to classify every course according to the nature of EfS integration. All these programmes have already worked to identify a statement about what makes a sustainable practitioner and how this is articulated in a graduate profile.

Simply by completing the matrix, either with a score of some kind or a comment, the extent of EfS within a programme quickly becomes clear, or areas for action are identified.

Programme element	Sustainability is there	It's there, but not explicit	It's not there but maybe should be	Sustainability is not really applicable
Whole course				
Graduate profile				
Learning outcomes				
Context				

In our experience, the "not really applicable" column accounts for only 5 to 10 percent of courses.

The model is good for identifying courses that should really have learning outcomes (for example, those that call for the application of knowledge), and those for which sustainability could provide some context for learning.

12.2 It seems such a big job to do it properly

Much of the EfS literature—indeed much of the higher education literature—argues for a transformation of education:

> Integrating aspects of sustainability cannot be realized without thinking very critically about the re-structuring of didactical arrangements. This re-orientation requires ample opportunity for staff members and students to embark on new ways of teaching and learning. (Wals & Jickling, 2002, p. 228)

Wals and Jickling write of the need to change from consumptive learning to discovery learning and creative problem solving; from teacher-centred to learner-centred arrangements; from knowledge accumulation to problem issue orientation. This is all very well, except if you've just been handed the responsibility of teaching 1,500 first years in three sittings of a 500-seat lecture theatre. Given this situation, should you throw up your hands in horror and declare that you can't possibly include sustainability? Of course not. The skill of a good lecturer is to engage every student as if individually; to find ways of making material relevant to the context of a group of students with cohorts from across the institution. This is not impossible, and sustainability may in fact make it easier: it is the built-in real-world context.

12.3 The professional development didn't help much

Professional development (PD) in this area is indeed useful. However, not having it should not prevent anyone from using their expertise of having lived in the world for quite a while:

> Sustainability capabilities will only be embedded into curriculum as part of a long lasting cultural change program, with a strong focus on well structured PD programs that allow for rigorous debate, discussion, sharing and learning in safe spaces within the academic community of universities. (Holdsworth, Wyborn, Bekessy, & Thomas, 2008, p. 135)

Disappointingly, PD sessions are often valued more for providing time away from the usual activities with the promise of a few free savouries or canapés than for actual professional enhancement. Scepticism about PD is widespread—and sometimes justifiable.

In relation to PD in EfS, I sometimes hear, "They had their heads in the clouds." It is entirely possible "they" actually did have their heads in the clouds and presented sustainability as touchy feely, woolly, waffly dogma, and did not present the case for EfS in a way that does justice to the need for a practical and ethical response to very real concerns. A deluge of good intentions and remonstrations is off-putting. Another possibility is that the person making this complaint is being contrarian and dislikes anything that smacks of tree hugging:

> Teaching about sustainability requires the transformation of mental models. Teaching about sustainability presupposes that those who teach consider themselves learners as well and that students and other concerned groups of interest are considered as repositories of knowledge and feelings too. Teaching about sustainability includes deep debate about normative, ethical and spiritual convictions and directly relates to questions about the destination of humankind and human responsibility. In this way it differs from a modernist and positivistic way of thinking. It incorporates notions of the possibility of the finiteness of human existence and trust in human creativity at the same time. (Wals & Jickling, 2002, p. 227)

12.4 Isn't it a bit hypocritical not to practise what we preach?

At times the lack of action on campus in support of sustainability acts as a barrier to including EfS in classrooms. There is a very understandable sense that until the institution starts composting waste or invests in a fleet of energy-efficient vehicles, it is not in a strong position to teach sustainability.

Teaching/learning and operations are closely linked. We have talked of the hidden curriculum, "walking the talk, and talking the walk", and seeing everything as a learning opportunity.

Given that a perfect state of sustainability is not possible, the best we can hope for is substantial progress along the continuum. Clearly we want to be able to demonstrate that we are doing the right things operationally, and wherever possible to use this as a learning opportunity. Some things, however, will take time to move significantly towards sustainability, and sometimes, despite best efforts, some clearly unsustainable thing remains intransigent.

The answer is, once again, to remind ourselves that sustainability is a journey, not a destination. We all have areas that need improvement, and recognising that is the first step. Perhaps an opportunity for learning is to prepare the business case for the new vehicles (or whatever it is that your students have noticed):

> It is criminally easy to accuse people of hypocrisy. But the accusation almost always misses the point, because by virtue of living in a carbon-based economy, none of us can say anything about emissions reduction without being hypocrites ourselves. Existing in the modern world creates carbon emissions. It's just a question of how bad a hypocrite you are. (Schendler, 2009, p. 84)

12.5 Where do I start? It isn't in the texts I use

Assuming that this statement is not simply another way of saying "This is not in my vocabulary", the issue of textbooks is quite real. Sustainability issues are being covered in some textbooks in the building industry, and it is certainly covered in texts for environmental law, ecology and other courses specific to the issue. However, many very useful texts are silent on the issue. What is needed here is to identify and compile supplementary material to augment other course material. A workbook might offer a supplementary chapter to a book, or interact with sections of a textbook with supplementary readings and questions. Departments can share materials and unit plans, which could help build up a workbook.

Mann, Muller, Davis, Roda and Young (2009) have described a framework for considering the scope and nature of textbooks for computing. The framework poses questions in six areas: philosophy; practice; practice guidelines; curriculum integration; linking and connection; and discipline-specific issues. The aim of the framework is to guide the assessment of existing resources towards establishing if and how the overall theme is appropriately addressed, and, if necessary, how the resource could be expanded or integrated with other resources to achieve appropriate theme coverage. In brief, this is what each area covers when applied to the discipline of computing:

- philosophy addresses the overarching ideas of computing for social and environmental sustainability
- practice addresses the issue of best sustainable practices within computing
- practice guidelines aim to ensure that the resource provides students with an insight on how to implement the specific practices—it calls for direct references to policies, standards, evaluation criteria, and in general practical steps and methods within the specific discipline
- curriculum integration deals with the level of integration of the resource with the existing computing curriculum
- linking and connecting are aimed at supporting the analysis of how well the resource links sustainable philosophy, practices and guidelines to the computing curriculum and the individual disciplines
- discipline specifics look at how well the resource fits a specific discipline while appropriately covering sustainability issues.

The authors' hope is that the framework will help computing educators to select the most appropriate resources for their courses.

12.6 I'm not confident that I have the expertise

EfS challenges thinking in education. It changes what it means to be an expert. With the objectives of transformative experiences for learners, the constructivist task really is about opening doors rather than filling vessels.

The first step is to examine your own beliefs. Think about what could be holding you back from gaining the expertise to teach sustainability. The beauty of EfS is that, as the teacher, you don't need to have the answers, only to encourage the questions. Learning to teach experientially, as a facilitator using participatory learning techniques, is the best way to deliver EfS:

> The role of the educational facilitator ... is crucial. There is no teaching or training involved. The field workers support and encourage (of course) but, more importantly, enable each group to take a critical relationship to both the accepted wisdom and to the process of learning. The workers' job is to listen actively and, through a process of critical questioning, help the group build a clear picture of what needs to be done. The field workers help the group assess the options and the risks of particular actions. They help people understand what the internet or other experts are telling them—and support the learners as they, small step by small step, move their project forward.
>
> So, how does this process fit any accepted definition of education? There is no teaching. No accepted wisdom passed down from professor to student. No single set of skills that, like a set of trusty spanners, can be made to fit each and every circumstance. (Fagan, 2009, p. 4)

Not knowing how to assess student progress is an issue that is often raised as a barrier to integrating EfS. In the long run, hopefully assessment already aligns with learning outcomes, which in turn reflect the graduate profile. So, address the graduate profile and work back from there. The majority of assessment for EfS will probably not be very different from the kinds of assessment already carried out in higher education (though not currently by all disciplines). Some EfS material will still be content led and be assessed with exams and essays; other EfS activities will be based around group work or work with stakeholders, field trips and internships, as well as self-reflection, all of which are already used in various disciplines and assessed by them in a range of ways.

Some educators might argue that assessment, as it exists, is inadequate and tests only the ability to regurgitate information rather than produce knowledge. One of the potential benefits of including EfS is that it is action based and intended to lead to modifications of systems or ways of doing things. It is not about being a passive receptacle, but about being an active facilitator of change. It is less about retention of information than about the production of knowledge.

12.7 I don't want to preach to the students

Preaching has no place in the classroom. Enthusiasm certainly does, particularly if combined with a lightness of touch whereby students are left to reach their own conclusions. Preachiness is never virtuous, even among preachers, because it is the incessant, unquestioning repetition of dogma. Preaching implies having an agenda, and that is not what EfS is about—other than the agenda to ensure the world remains habitable for our descendants. EfS seeks to correct society's somnambulist walk towards environmental decrepitude and prompts students to ask, "What consequences do my actions have? If there are potentially negative consequences can I (and should I) work to ameliorate them?"

The problem is how to cover the degradation parts in a positive way. We are trying to do two things that are not happy bedfellows: we are striving to be both hard hitting and encouraging (or at least not preachy). We are, in essence, delivering a doomsday prophecy and we need to offset this with a way forward, and, frankly, lighten up. Being preachy and earnest makes environmentalists the target of humour. Instead of being the targets, we should be using humour as a tool to engage and as a tension-release mechanism. I'm an advocate of incidental humour—the wry observation. This, of course, is hard to find. Rustle the Leaf (http://www.rustletheleaf.com/lessonplans.html) does a great job of incorporating a lighter approach to the environment, and helpfully produces lesson plans which meet both objectives, although some of it is designed for children.

So here's a challenge for anyone who enjoys teaching. We need to present sustainability as an opportunity, not as imminent disaster. We've all had experiences of feeling overwhelmed, or seeing others overwhelmed, by the enormousness of the issues and feeling that whatever small contribution we can make is so slight that we might as well not bother. It is this attitude that we must be careful not to engender in our students. Rather, we must be positive and action based.

12.8 How do I accommodate all the different views and perspectives?

Every class will include a diversity of views on EfS issues, and this can be challenging. However, it can also be an opportunity rather than a barrier. Use the diversity of viewpoints to your advantage:

> Effective policy for a society depends on successful creation of shared meaning among—not within—cultural groups. Two points are particularly relevant here: (1) shared social values as well as shared meanings are created through effective social interaction, they do not exist a priori, nor are they merely the intersection of individual values, instead, they are created through social interaction; and (2) the only effective way to achieve this kind of interaction is through open dialogue, where social learning is necessary for forming as well as for changing shared social values. (Stagl, 2007, p. 55)

Department for Environment, Food and Rural Affairs (Defra) (2008) has done much work on different attitudes and behaviours. It uses a segmentation tool to explore different groups in the population (see Figure 10). The same groups may be present in your class. How might you engage each group?

Brown (2005) has argued that transforming a person's world view in relation to the environment can be a difficult process, and advocates a process of segmenting your audience to translate sustainability messages into the different world views of the population. An

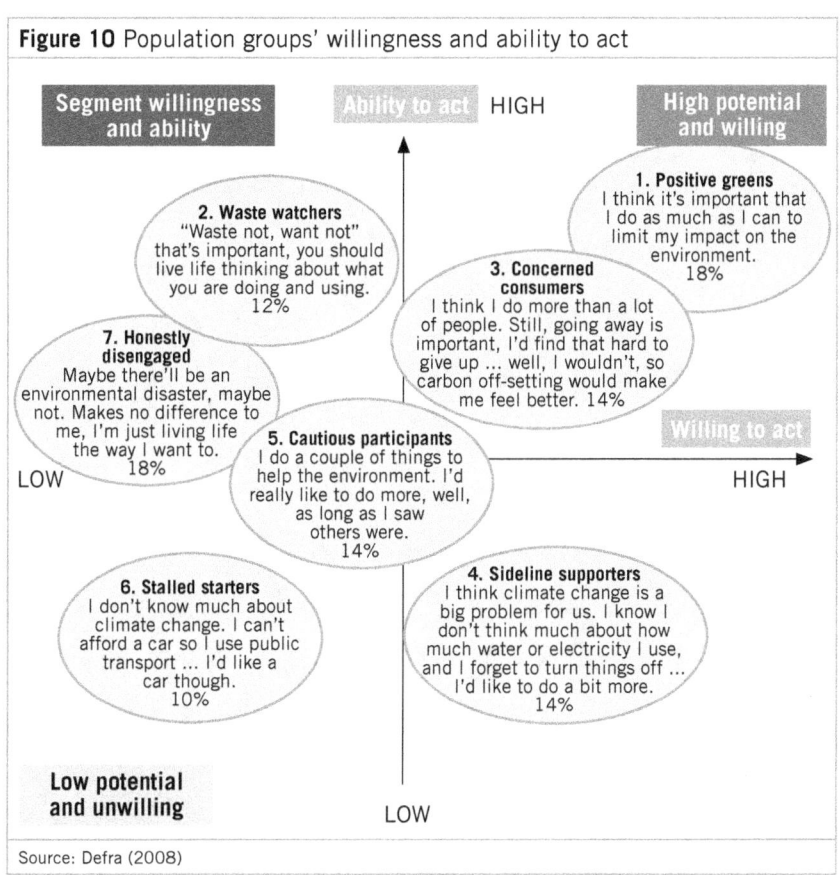

Figure 10 Population groups' willingness and ability to act

Source: Defra (2008)

"eco-manager", for example, who has a stewardship ethos, may respond positively to concepts such as duty, honour, country, good citizenship and so on. An "eco-strategist", on the other hand, has a rational ethos and sees "nature for profit and play"—something to be managed, and may respond to an appeal to competitive advantage, drawing on proven success and returns.

Conclusion

Becoming sustainable is a challenge that faces all of us. I'm an optimist—I believe that we *can* take up the challenge and find ways to become more sustainable as a society.

As educators, our greatest impact can be through preparing our students to handle the sustainability issues ahead of them. In this book I've shown how Otago Polytechnic has set out on the journey to integrating sustainability into all our programmes—and some ways to recognise and help break down barriers along the way. Together we are making progress towards a future in which every graduate becomes a sustainable practitioner. Please join us.

References

Adomssent, M., Godemann, J., & Michelsen, G. (2007). Transferability of approaches to sustainable development at universities as a challenge. *International Journal of Sustainability in Higher Education, 8*(4), 385–402.

Alvarez, A., & Rogers, J. (2006). Going "out there": Learning about sustainability in place. *International Journal of Sustainability in Higher Education, 7*(2), 176–188.

AtKisson, A. (2008). *The ISIS agreement: How sustainability can improve organizational performance and transform the world.* London: Earthscan.

Bair, B., & Cohoon, J. M. (Eds.). (2004). Special issue on gender-balancing computing education. *Journal on Educational Resources in Computing, 4*(1), 1. doi: 10.1145/1060071.1060072

Barnbrook, J. (2007). *Barnbrook Bible, the graphic design of Jonathon Barnbrook: Booth-Clibborn Editions.* London: Barnbrook.

Bartlett, J. (Ed.). (1980). *Familiar quotations: A collection of passages, phrases and proverbs traced to their sources in ancient and modern literature* (15th ed. rev.). London: Macmillan.

Bekessy, S. A., Samson, K., & Clarkson, R. E. (2007). The failure of non-binding declarations to achieve university sustainability. *International Journal of Sustainability in Higher Education, 8*(3), 301–316.

Benford, S., Giannachi, G., Koleva, B., & Rodden, T. (2009). From interaction to trajectories: Designing coherent journeys through user experiences.

REFERENCES

Proceedings of the 27th International Conference on Human Factors in Computing Systems. Boston, MA: Association for Computing Machinery.

Benyo, J., White, J., Ross, J., Wiehe, B., & Sigur, M. (2009). *New image for computing: Report on market research April 2009.* Retrieved from Association for Computing Machinery: http://www.acm.org/press-room/membership/NIC.pdf

Berry, T. (2000). *The great work.* New York: Harmony Books.

Berry, T. (2009). *The sacred universe: Earth, spirituality, and religion in the twenty-first century.* New York: Columbia University Press.

Berry, W. (2008). The way of ignorance. In W. Vitek, B. Vitek, & W. Jackson (Eds.), *The virtues of ignorance: Complexity, sustainability, and the limits of knowledge* (pp. 37–50). Lexington, KY: University Press of Kentucky.

Blevis, E. (2006). Advancing sustainable interaction design: Two perspectives on material effects. *Design Philosophy Papers,* (4), 1–25.

Blewitt, J. (2004). Sustainability and lifelong learning. In J. Blewitt & C. Cullingford (Eds.), *The sustainability curriculum: The challenge for higher education* (pp. 24–42). London: Earthscan.

Blewitt, J. (2009). New media literacy: Communication for sustainability. In A. Stibbe (Ed.), *The handbook of sustainability literacy: Skills for a changing world* (pp. 111–116). Totnes, UK: Green Books.

Bosselmann, K. (2008). *The principle of sustainability: Transforming law and governance.* Farnham, UK: Ashgate Publishing.

Brown, B. C. (2005). Integral communications for sustainability. *Kosmos,* 4(2), 17–22.

Carrithers, D. F., & Peterson, D. (2006). Conflicting views of markets and economic justice: Implications for student learning. *Journal of Business Ethics,* 69(4), 373–387.

Carson, R. (1962). *Silent spring.* Boston, MA: Houghton Mifflin.

Charpentier, C. (1994). *Barriers to environmental education in Costa Rican state universities: Theory, analysis and recommendations for intervention programs.* Unpublished doctoral thesis, University of Idaho, Moscow, ID.

Computing Research Association. (1974–2010). [Taulbee survey results.] Available from Computing Research Assocation: http://www.cra.org/resources/taulbee/2007-2008_taulbee_survey/ 5/8/2010

Corcoran, P. B., & Wals, A. E. J. (2004). The problematics of sustainability in higher education: An introduction. In P. B. Corcoran & E. J. W. Arjen (Eds.), *Higher education and the challenge of sustainability: Problematics, promise and practice* (pp. 3–6). New York: Kluwer Academic.

REFERENCES

Cortese, A. D. (1999). *Education for sustainability: The need for a new human perspective.* Retrieved from http://www.secondnature.org/pdf/snwritings/articles/humanpersp.pdf

Cortese, A. D., & McDonough, W. (2001). *Education for sustainability: Accelerating the transition to sustainability through higher education.* Retrieved from http://www.secondnature.org/pdf/snwritings/articles/AccTheTrans.pdf

Costanza, R. (1991). *Ecological economics: The science and management of sustainability.* New York: Columbia University Press.

Costanza, R., & Wainger, L. (1991). *Ecological economics: Mending the earth.* Berkeley, CA: North Atlantic Books.

von Cotta, H. (1817/2000). *Anweisung zum Waldbau.* Translated in *Forestry Quarterly*, Vol. 1, 1902/1903. Preface reprinted in *Forest History Today*, Fall, 2000, 27–28.

Cruess, R., Cruess, S., & Johnston, S. (1997). Teaching medicine as a profession in the service of healing. *Academic Medicine, 72,* 942–952.

Cruess, R. L., & Cruess, S. R. (2008). Expectations and obligations: Professionalism and medicine's social contract with society. *Perspectives in Biology and Medicine, 51*(4), 579–598.

Crutzen, P. J. (2002). Geology of mankind. *Nature, 415*(6867), 23.

Crutzen, P. J. (2005). Human impact on climate has made this the "Anthropocene Age". *New Perspectives Quarterly, 22*(2), 14–16.

Cullingford, C. (2004). Sustainability and higher education. In J. Blewitt & C. Cullingford (Eds.), *The sustainability curriculum: The challenge for higher education* (pp. 13–23). London: Earthscan.

Daly, H. E. (1996). *Beyond growth: The economics of sustainable development.* Boston, MA: Beacon Press.

van Dam-Mieras, R. (2006). Learning for sustainable development: Is it possible within the established higher education structures? In J. Holmberg & B. E. Samuelsson (Eds.), *Drivers and barriers for implementing sustainable development in higher education* (pp. 13–18). Paris: UNESCO Education.

Datschefski, E. (n.d.). *The total beauty of sustainable products.* Retrieved from http://www.biothinking.com/btintro.htm

Davies, K. (2009). A learning society. In A. Stibbe (Ed.), *The handbook of sustainability literacy: Skills for a changing world* (pp. 215–221). Totnes, UK: Green Books.

Davies, L. (2007). *Ecological midwifery.* Retrieved 23 May 2008, from http://www.withwoman.co.uk/contents/info/ecomid.html

REFERENCES

Department for Environment, Food and Rural Affairs. (2008). *A framework for pro-environmental behaviours*. London: Author.

Design Council. (2008). *The good design plan*. London: Author.

The Designers Accord. (n.d.,a). *Mission*. Retrieved 23 February 2011, from http://www.designersaccord.org/mission/

The Designers Accord. (n.d.,b). *Join us*. Retrieved 23 February 2011, from http://www.designersaccord.org/join-us/

Diamond, J. (2005). *Collapse: How societies choose to fail or succeed*. New York: Viking.

Downing, P., Cox, J., & Fell, D. (2004). *Liveability vs sustainability: Bad habits & hard choices*. London: Office of the Deputy Prime Minister. Available from http://www.communities.gov.uk/publications/corporate/livabilityvsustainability

Dumaine, B. (2008). *The plot to save the planet: How visionary entrepreneurs and corporate Titans are creating real solutions to global warming*. New York: Crown Business.

Earth Charter International. (2009a). *Earth Charter*. Retrieved from http://www.earthcharterinaction.org/content/pages/Read-the-Charter.html

Earth Charter International. (2009b). *A guide for using the Earth Charter in education*. Retrieved from http://www.earthcharterinaction.org/invent/images/uploads/EC_Education_Guide_2%20APRIL_2009.pdf

Eisen, A., & Barlett, P. (2006). The Piedmont project: Fostering faculty development toward sustainability. *Journal of Environmental Education, 38*(1), 25–36.

Evetts, J. (2003). The construction of professionalism in new and existing occupational contexts: Promoting and facilitating occupational change. *The International Journal of Sociology and Social Policy, 23*(4/5), 22–35.

Fagan, G. (2009). The emerging paradigm. In A. Stibbe (Ed.), *The handbook of sustainability literacy: Skills for a changing world*. Retrieved 8 March 2011, from http://arts.brighton.ac.uk/__data/assets/pdf_file/0018/6246/The-Emerging-Paradigm.pdf

Ferrer-Balas, D., Adachi, J., Banas, S., Davidson, C. I., Hoshikoshi, A., Mishra, A., et al. (2008). An international comparative analysis of sustainability transformation across seven universities. *International Journal of Sustainability in Higher Education, 9*(3), 295–316.

Fish, S. (2008). *Save the world on your own time*. Oxford: Oxford University Press.

Fox, D. (2009). An overarching framework for sustainability. *Built Environment, 35*(3), 302–307.

REFERENCES

Franz-Balsen, A., & Heinrichs, H. (2007). Managing sustainability communication on campus: Experiences from Lüneburg. *International Journal of Sustainability in Higher Education, 8*(4), 431–445.

Freire, P., & Freire, A. M. A. (2000). *Pedagogy of the heart.* New York: Continuum International Publishing.

Fuller, R. B. (1969). *Operating manual for Spaceship Earth.* Carbondale, IL: Southern Illinois University Press.

Gardner, H. (1993). *Frames of mind: The theory of multiple intelligences* (2nd ed.). New York: Basic Books.

Global Action Plan. (2007). *An inefficient truth.* London: Author.

Hallstedt, S. (2008). *A foundation for sustainable product development.* Unpublished doctoral thesis, School of Engineering, Department of Mechanical Engineering, Blekinge Institute of Technology, Karlskrona, Sweden.

Hamilton, C., & Denniss, R. (2005). *Affluenza: When too much is never enough.* Crows Nest, NSW: Allen & Unwin.

Healey, M. (2009). An inconvenient data center. *Information Week.* Retrieved 19 January 2009, from http://www.informationweek.com/news/infrastructure/showArticle.jhtml?articleID=212900868

Herrmann, M. (2007). The practice of sustainable education through a participatory and holistic teaching approach. *Communication, Cooperation, and Participation for a Sustainable Future, 1,* 72–87.

The Higher Education Academy. (2006). *Sustainable development in higher education: Current practice and future developments—A progress report for senior managers in higher education.* York, UK: Author. Available at http://www.heacademy.ac.uk/assets/York/documents/resources/resourcedatabase/id587_sustainable_development_managers_report.pdf

Holdsworth, S., Wyborn, C., Bekessy, S., & Thomas, I. (2008). Professional development for education for sustainability. *International Journal of Sustainability in Higher Education, 9*(2), 131–146.

Holland, D. K. (2001). *Design issues: How graphic design informs society.* New York: Allworth Press.

Holmberg, J., & Samuelsson, B. E. (2006). Drivers and barriers for implementing sustainable development in higher education: Executive summary. In J. Holmberg & B. E. Samuelsson (Eds.), *Drivers and barriers for implementing sustainable development in higher education* (pp. 7–11). Paris: UNESCO Education.

IBM & New Zealand Business Council for Sustainable Development. (2008). *The environment, business and technology.* Wellington: Author.

REFERENCES

Industrial Designers Society of America. (2001). *IDSA recognizes the following ecological principles*. Retrieved 13 January 2011, from http://www.idsa.org/content/content1/idsa-recognizes-following-ecological-principles

International Union for Conservation of Nature, United Nations Environment Programme & World Wildlife Fund. (1991). *Caring for the Earth*. Gland, Switzerland: International Union for Conservation of Nature.

James, P., & Hopkinson, L. (2009). *Sustainable ICT in further and higher education*. Bradford, UK: Joint Information Services Committee.

Jensen, B. B., & Schnack, K. (1997). The action competence approach in environmental education. *Environmental Education Research, 3*(2), 163–179.

Koehler, A., & Som, C. (2005). Effects of pervasive computing on sustainable development. *Technology and Society Magazine, 24*(1), 15–23.

Koutsouris, A. (2009). Sustainability, crossdisciplinarity and higher education: An agronomic point of view. *US–China Education Review, 6*(3), 13–27.

Lambert, M. J. (2004). *Bergin and Garfield's handbook of psychotherapy and behavior change*. New York: Wiley.

Lautensach, A. K. (2004, November–December). *A tertiary curriculum for sustainability*. Paper presented at the Australian Association of Research in Education conference, Melbourne. Retrieved from http://www.aare.edu.au/04pap/lau04260.pdf

Love, K., & Love, D. (2008, July). *"What do we do now?" Teaching sustainability in IT*. Paper presented at the 17th annual conference of the National Advisory Committee on Computing Qualifications, Auckland.

Makower, J. (2009). *State of green business 2009*. Oakland, CA: Greener World Media.

Mann, S. (2008). *Sustainable practitioner*. Unpublished PowerPoint presentation.

Mann, S., & Buissink-Smith, N. (2000, June). *What the students learn: Learning through empowerment*. Paper presented at the 13th annual conference of the National Advisory Committee on Computing Qualifications, Wellington.

Mann, S., & Ellwood, K. (Eds.). (2009). *A simple pledge: Towards sustainable practice*. Dunedin: Otago Polytechnic.

Mann, S., Muller, L., Davis, J., Roda, C., & Young, A. (2009). Computing and sustainability: Evaluating resources for educators. *ACM SIGCSE Bulletin, 41*(4), 144–155. doi: 10.1145/1709424.1709459

Mann, S., Russell, K., Camp, J., Crook, M., & Wikaira, J. (2006, July). *Māori game design*. Paper presented at the 19th annual conference of the National Advisory Committee on Computing Qualifications, Wellington.

REFERENCES

Mann, S., & Smith, L. (2006, July). *Arriving at an agile framework for teaching software engineering*. Paper presented at the 19th annual conference of the National Advisory Committee on Computing Qualifications, Wellington.

Mann, S., Smith, L., Shephard, K., Smith, N., & Deaker, L. (2009, July). *Benchmarking sustainability values of incoming computing students*. Paper presented at the 22nd annual conference of the National Advisory Committee on Computing Qualifications, Napier.

Marsh, G. P. (1864). *Man and nature; or physical geography as modified by human action*. New York: Charles Scribner.

Martin, S. (2005). Sustainability, systems thinking and professional practice. *Systemic Practice and Action Research, 18*(2), 163–171.

McDonald, P., & Lassoie, J. P. (1996). *The literature of forestry and agroforestry*. Ithaca, NY: Cornell University Press.

McDonough, W., & Braungart, M. (2002). *Cradle to cradle: Remaking the way we make things*. San Francisco: North Point Press.

McKeown, R. (2002). *Education for sustainable development toolkit*. Retrieved 15 March 2007, from http://www.esdtoolkit.org

Meadows, D. H., Meadows, D. L., Randers, J., & Behrens, W. W. (1972). *The limits to growth: A report for the Club of Rome's project on the predicament of mankind*. New York: Universe Books.

Mellalieu, P. (2009, April). *Shifting frontiers, new priorities, creating pathways: Elevating the case for tertiary education for sustainable development in New Zealand*. Paper presented at the New Zealand Tertiary Education Summit, Wellington.

Millenium Ecosystem Assessment. (2005). *Ecosystems and human well-being: Opportunities and challenges for business and industry*. Washington, DC: World Resources Institute.

Miller, W. R., & Rollnick, S. (2002). *Motivational interviewing: Preparing people for change* (2nd ed.). New York: The Guildford Press.

Milne, M. J., Kearins, K., & Walton, S. (2006). Creating adventures in wonderland: The journey metaphor and environmental sustainability. *Organization, 13*(6), 801–839. doi: 10.1177/1350508406068506

Morris, D., & Martin, S. (2009). Complexity, systems thinking and practice: Skills and techniques for managing complex systems. In A. Stibbe (Ed.), *The handbook of sustainability literacy: Skills for a changing world* (pp. 156–164). Totnes, UK: Green Books.

Mulder, K. F., & Jansen, J. L. A. (2006). Integrating sustainable development in engineering education: Reshaping university education by organizational learning. In J. Holmberg & B. E. Samuelsson (Eds.), *Drivers and barriers for implementing sustainable development in higher education* (pp. 69–74). Paris: UNESCO Education.

National Advisory Committee on Computing Qualifications. (2007, July). National Advisory Committee on Computing Qualifications annual general meeting minutes.

New Zealand Government. (2005). *The Digital Strategy: Creating our digital future.* Retrieved from http://www.digitalstrategy.govt.nz

Newman, J. (2009). Values reflection and the Earth Charter. In A. Stibbe (Ed.), *The handbook of sustainability literacy: Skills for a changing world* (pp. 99–104). Totnes, UK: Green Books.

Nicolaides, A. (2006). The implementation of environmental management towards sustainable universities and education for sustainable development as an ethical imperative. *International Journal of Sustainability in Higher Education, 7*(4), 414–424.

Nursing Council of New Zealand. (2009). *Competencies for registered nurses.* Retrieved from http://www.nursingcouncil.org.nz/download/73/rn-comp.pdf

Nussbaum, E. (2009, January). The new journalism: Goosing The Gray Lady. *New York Magazine.* Retrieved 14 January 2009, from http://nymag.com/news/features/all-new/53344/

Organisation for Economic Co-operation and Development. (2009). *Green at fifteen? How 15-year-olds perform in environmental science and geoscience in PISA 2006.* Paris: OECD Publications.

Orr, D. W. (1992). *Ecological literacy: Education and the transition to a postmodern world.* Albany, NY: State University of New York Press.

Orr, D. W. (1994). *Earth in mind: On education, environment, and the human prospect.* Washington, DC: Island Press.

Osbaldiston, R., & Sheldon, K. M. (2002). Social dilemmas and sustainable development: Promoting the motivation to "cooperate with the future". In P. Schmuck & W. Schultz (Eds.), *The psychology of sustainability* (pp. 37–58). Boston, MA: Kluwer.

Otago Polytechnic. (2007). *Education for sustainability at Otago Polytechnic.* Retrieved 21 February 2011, from http://www.otagopolytechnic.ac.nz/about/sustainable-practice/education-for-sustainability.html

Otago Polytechnic. (2008). *Programme document: 11009.20 ND0769 National Diploma in Hospitality (Management) (Level 5) and OT4708 Certificate in Professional Restaurant, Bar and Wine (Level 4). Version 2. May 2008.* Unpublished document.

REFERENCES

Otago Polytechnic. (2009a). *Programme document: NC1014 National Certificate in Horticulture (Level 4). Version 1e. 21 October 2009.* Unpublished document.

Otago Polytechnic. (2009b). *Programme document: Bachelor of Design (Specialty) Version 2. September 2009.* Unpublished document.

Otago Polytechnic. (2009c). *Programme document: OT4728 Bachelor of Nursing. Version 4. August 2009.* Unpublished document.

Otago Polytechnic. (2009d). *Academic Board paper A148/09 Framework for inclusion of core capabilities in programmes.* Unpublished document.

Parliamentary Commissioner for the Environment. (2004). *See change: Learning and education for sustainability.* Wellington: Office of the Parliamentary Commisioner for the Environment.

Paten, C. J. K., Palousis, N., Hargroves, K., & Smith, M. (2005). Engineering sustainable solutions program: Critical literacies for engineers portfolio. *International Journal of Sustainability in Higher Education, 6*(3), 265–277.

van Peborgh, E., & Odiseo Team. (2008). *Sustainability 2.0: Networking enterprises and citizens to face world challenges.* Retrieved from http://sustainabilitythebook.com/SustainabilityTheBook.pdf

Phillips, A. (2009). Institutional transformation. In A. Stibbe (Ed.), *The handbook of sustainability literacy: Skills for a changing world* (pp. 209–214). Totnes, UK: Green Books.

Placet, M., Anderson, R., & Fowler, K. M. (2005). Strategies for sustainability. *Research Technology Management, 48*(5), 32–41.

Polistina, K. (2009). Cultural literacy: Understanding and respect for the cultural aspects of sustainability. In A. Stibbe (Ed.), *The handbook of sustainability literacy: Skills for a changing world* (pp. 117–123). Totnes, UK: Green Books.

Rice, V. J., & Duncan, J. R. (2006). *What does it mean to be a "professional" ... and what does it mean to be an ergonomics professional? A position paper sponsored by the Foundation for Professional Ergonomics.* Retrieved from http://www.ergofoundation.org/FPE1_Professionalism.pdf

Robinson, H. (1994). *The ethnography of empowerment: The transformative power of classroom interaction.* London: Falmer Press.

Robinson, Z. (2009). Greening business: The ability to drive environmental and sustainability improvements in the workplace. In A. Stibbe (Ed.), *The handbook of sustainability literacy: Skills for a changing world* (pp. 130–136). Totnes, UK: Green Books.

Schendler, A. (2009). *Getting green done: Hard truths from the front lines of the sustainability revolution.* New York: Public Affairs.

REFERENCES

Scott, W., & Gough, S. (2006). Universities and sustainable development in a liberal democracy: A reflection on the necessity for barriers to change. In J. Holmberg & B. E. Samuelsson (Eds.), *Drivers and barriers for implementing sustainable development in higher education* (pp. 80–95). Paris: UNESCO Education.

Second Nature. (n.d.). *Sustainability curriculum framework*. Retrieved 28 February 2011, from http://www.secondnature.org/pdf/snwritings/factsheets/framework.pdf

Senge, P. M., Laur, J., Schley, S., & Smith, B. (2006). *Learning for sustainability*. Cambridge, MA: Society for Organisational Learning.

Senge, P. M., Smith, B., Schley, S., Laur, J., & Kruschwitz, N. (2008). *The necessary revolution: How individuals and organizations are working together to create a sustainable world*. New York: Doubleday Publishing.

Shephard, K., Mann, S., Smith, N., & Deaker, L. (2009). Benchmarking the environmental values and attitudes of students in New Zealand's post-compulsory education. *Environmental Education Research, 15,* 571–587.

Sibbel, A. (2009). Pathways towards sustainability through higher education. *International Journal of Sustainability in Higher Education, 10*(1), 68–82.

Smith, L., Mann, S., & Buissink-Smith, N. (2001). Crashing a bus full of empowered software engineering students. *New Zealand Journal of Applied Computing and Information Technology, 5*(2), 69–74.

Solis, B., & Breakenridge, D. (2009). *Putting the public back in public relations*. Upper Saddle River, NJ: Pearson Education.

Springett, D., & Kearins, K. (2005). Educating for sustainability: An imperative for action. *Business Strategy and the Environment, 14*(3), 143–145.

Stagl, S. (2007). Theoretical foundations of learning processes for sustainable development. *International Journal of Sustainable Development and World Ecology, 14,* 52–62.

Steffen, W., Crutzen, P., & McNeill, J. (2007). The Anthropocene: Are humans now overwhelming the great forces of nature? *Ambio, 36*(8), 614–621.

Stephens, J. C., Hernandez, M. E., Roman, M., Graham, A. C., & Scholz, R. W. (2008). Higher education as a change agent for sustainability in different cultures and contexts. *International Journal of Sustainability in Higher Education, 9*(3), 317–338.

Sterling, S. (2004a). An analysis of the development of sustainability education internationally: Evolution, interpretation and transformative potential. In J. Blewitt & C. Cullingford (Eds.), *The sustainability curriculum: The challenge for higher education* (pp. 43–62). London: Earthscan.

REFERENCES

Sterling, S. (2004b). Higher education, sustainability, and the role of systemic learning. In P. B. Corcoran & E. J. W. Arjen (Eds.), *Higher education and the challenge of sustainability: Problematics, promise and practice* (pp. 50–70). New York: Kluwer Academic.

Sterling, S. (2009). Ecological intelligence: Viewing the world relationally. In A. Stibbe (Ed.), *The handbook of sustainability literacy: Skills for a changing world* (pp. 77–83). Totnes, UK: Green Books.

Strachan, G. (2009). Systems thinking: The ability to recognize and analyse the inter-connections within and between systems. In A. Stibbe (Ed.), *The handbook of sustainability literacy: Skills for a changing world* (pp. 84–88). Totnes, UK: Green Books.

Svanström, M., Lozano-García, F. J., & Rowe, D. (2008). Learning outcomes for sustainable development in higher education. *International Journal of Sustainability in Higher Education, 9*(3), 339–351.

Tertiary Education Commission. (2007). *Tertiary Education Strategy 2007–2012 incorporating the Statement of Tertiary Education Priorities 2008–2010*. Retrieved from http://www.minedu.govt.nz/index.cfm?layout=document&documentid=11727&data=1

Tomkinson, B. (2009). Coping with complexity: The ability to manage complex sustainability problems. In A. Stibbe (Ed.), *The handbook of sustainability literacy: Skills for a changing world* (pp. 165–170). Totnes, UK: Green Books.

Tormey, R., Liddy, M., & Hogan, D. (2009). Interdisciplinary literacy: The ability to critique disciplinary cultures and work effectively across disciplines. In A. Stibbe (Ed.), *The Handbook of Sustainability Literacy: Skills for a Changing World*. Retrieved 18 March 2011, from http://arts.brighton.ac.uk/stibbe-handbook-of-sustainability/additional-chapters/interdisciplinary-literacy

United Nations. (1992). *Earth Summit: Agenda 21: The United Nations Programme of Action from Rio: "The Rio Declaration on Environment and Development"*. Retrieved 18 March 2011, from http://www.un.org/esa/dsd/agenda21/res_agenda21_36.shtml

United Nations. (2004). *United Nations decade of education for sustainable development: Information brief: Higher education*. Retrieved 15 March 2007, from http://www.unesco.org/education/tlsf/TLSF/decade/img/DESDbrief.pdf

United Nations Educational, Scientific and Cultural Organization. (2006). *United Nations Decade of Education for Sustainable Development (2005–2014)*. Retrieved 19 March 2011, from http://portal.unesco.org/science/en/ev.php-URL_ID=4931&URL_DO=DO_TOPIC&URL_SECTION=201.html

Vallero, D. A. (2005). *Paradigms lost: Learning from environmental mistakes, mishaps and misdeeds*. Amsterdam: Butterworth-Heinemann.

REFERENCES

Velazquez, L., Munguia, N., & Sanchez, M. (2005). Deterring sustainability in higher education institutions: An appraisal of the factors which influence sustainability in higher education institutions. *International Journal of Sustainability in Higher Education*, 6(4), 383–391.

Wals, A. E. J., & Jickling, B. (2002). "Sustainability" in higher education: From doublethink and newspeak to critical thinking and meaningful learning. *International Journal of Sustainability in Higher Education*, 3(3), 221–232.

Welsh Assembly Government. (2007). *A common understanding for education for sustainable development and global citizenship (ESDGC) in Wales*. Retrieved from http://www.glos.ac.uk/research/iris/strands/esd/Documents/esdgcFinalEng1.pdf

Welsh Assembly Government. (2008). *Education for sustainable development and global citizenship: Information for teacher trainees and new teachers in Wales*. Retrieved from http://www.esd-wales.org.uk/english/higher_ed/downloads/ESDGC%20teacher%20trainees%20%28e%29.pdf

Williams, P. M. (2008). *University leadership for sustainability: An active dendritic framework for enabling connection and collaboration*. Unpublished doctoral thesis, Victoria University of Wellington.

World Commission on Environment and Development. (1987). *Our common future* [Brundtland report]. Retrieved 5 June 2010, from http://www.un-documents.net/wced-ocf.htm

Wright, T. S. A. (2004). The evolution of sustainability declarations in higher education. In P. B. Corcoran & E. J. W. Arjen (Eds.), *Higher education and the challenge of sustainability: Problematics, promise and practice* (pp. 7–20). New York: Kluwer Academic.

Zelezny, L. C., & Schultz, P. W. (2000). Promoting environmentalism. *Journal of Social Issues*, 56(3), 365–371.

The Beeby Fellowship

The Beeby Fellowship is a joint initiative of the New Zealand Council for Educational Research (NZCER) and the New Zealand Commission for UNESCO. Dr Clarence Beeby was the inaugural director of NZCER in 1934 and was Assistant Director-General of UNESCO, 1948–9. He died in March 1998, aged 95.

The purpose of the Fellowship is to enable a person actively involved in an innovative programme or school to document, analyse and write a resource about the programme. The focus of the Fellowship is on innovation that will enhance teaching practice and students' learning.

Samuel Mann was awarded the Beeby Fellowship in 2009 to write *The Green Graduate: Educating Every Student as a Sustainable Practitioner*.

www.ingramcontent.com/pod-product-compliance
Lightning Source LLC
Chambersburg PA
CBHW080635230426
43663CB00016B/2879